C0-AKV-459

Technology
Fire in a Dark World

To the women in my life–

a mother for her gift of optimism,
a wife for her confidence in me,
two daughters for proof of the better
world ahead

Preface

Outside my window of the 747, only the simple silhouette of a wing interrupted the predawn blue. Then bright yellow flares pierced the void below. "Oil rigs," said the proud Norwegian next to me. Platforms, stealing oil from beneath the North Sea, were burning waste gases.

In my comfortable man-made environment at 37,000 feet, I pictured rugged workers on the man-made platforms barely topping the waves. They and the men piloting my plane are masters of other worlds. I am not at home on an oil rig or the flight deck of a plane. They and their technologies enable me to travel, however. Although we do not see each other, we are linked together by technology.

Developed and used by brave creative people, technological progress has served me well. Yet, not everyone feels that way. "What is technological progress?" they ask. "Just an endless giving way to change? Are we better off because of it? Are *we* any better?"

People are divided over these questions. They tend to believe either that technology has caused most of man's problems or that it can solve most of them. While they debate, a number of signs have begun to indicate that the pace of progress is slowing.

The confusion over whether technology is good or bad reflects our lack of understanding of what people want from life and why technology troubles them. We live in a technological society which, at the same time, is struggling to be more humane.

Some people feel we are accelerating ourselves into oblivion, but the development of new scientific knowledge and applications of fundamentally new technology is withering. Doomsday talk about the evils of technology should be countered with alarm over the troubled state of our technological capability.

We waste our time debating whether technology is good or bad. We should be asking the right questions: What forms of technology do we want? To what ends should we apply them? Who will determine what is desirable and what is safe? How can we maximize our technological contributions for a better life?

We should constantly question the users of technology. Challenging its very existence is wrong, however, because the better life the whole world wants depends on appropriately used technology. The mysteries of our origin and our destiny compel us to continue learning and innovating.

We cannot go back. The question is: Which way is forward? Only by blending the values of the humanist and the technologist can we ensure that the "fire" of inquiry and innovation neither goes out—nor goes out of control.

We can begin by surveying the fruits of technology—the bitter and the sweet—and exploring the many aspects of the role technology has played in the history of man. By no means an exhaustive study of each aspect, this book is intended to stimulate thinking by people on all sides of the technology debate. At the risk of oversimplifying the issues, I hope to promote an awareness of complex relationships and bring more people into the discussion of the right questions. I do not presume to have detailed prescriptions for all the ethical, moral, political, and economic problems posed by technology. I can only share my positive view of man and his potential for building a better world.

PERRY PASCARELLA
Bay Village, Ohio

Acknowledgments

My concern for this topic may have begun with a half day spent with Donald Alstadt, president of Lord Corporation, as we discussed all sorts of economic, social, and political subjects nearly three years ago. During the following year, it grew on comments made during interviews for various stories, casual discussions with friends, and personal reading.

When I told Stan Modic, the editor of *Industry Week*, that I wanted to dig into this topic for a story and possibly even a book, he responded with full support. That began another year in which countless people sat for long interviews—most of them ending with offers of any further help I needed. As I described my story idea and posed my questions, many remarked, "you really ought to do a book on this." The article "Goodbye technology, farewell future" appeared in the September 12, 1977 issue of *IW*. But the work on the book went on. By now, nearly any conference I attended on any subject, anyone I talked with, or anything I read was apt to add a piece here and there for a book.

When all the pieces finally come together and you find you have a book, how can you begin to thank all those who have helped with written and spoken words? Many, but not all, of the people I interviewed are mentioned in this book. Their affiliations and titles reflect where they were when I caught them; some have moved to new positions since then. The thoughts of writers spanning many years helped shape this book. I thank the publishers who have granted permission to borrow from the literature which brings so much light to bear on this subject. The literature has been supplemented through personal interviews with some of these writers; therefore, the reader will find no reference in the chapter notes for some of their comments.

Special thanks go to my colleagues on the staff of *Industry Week* for sharing their information and their social concern and, particularly, for their encouragement.

There is only one person on whom completion of this project totally depended—my wife. We shared many hours of loneliness to make this a reality.

Contents

Preface vii
Acknowledgments ix

PART I TOWARD A BETTER LIFE

1 Fire and Freedom 3
2 Less Innovation, Limited Options 13
3 Growing List of Wants 26

PART II MISCONCEPTIONS AND APPREHENSIONS

4 Broken Promises 39
5 Too Much Growth 47
6 The Perception of Change 55
7 Planet in Peril 62
8 Inhumane Systems 71
9 Destroyer of Jobs 77

PART III THE WORSENING CLIMATE FOR INNOVATION

10 Overreaction by Overregulation 89
11 Flickering Corporate Spirit 101

PART IV THE DEMOCRATIZATION OF TECHNOLOGY

12 The People Will Decide 119
13 Private Enterprise and Public Goals 133
14 Dialogue and Control 147
15 If Man Is to Be Man 157

Chapter Notes 163
Bibliography 169

Technology
Fire in a Dark World

PART I
Toward A Better Life

1 Fire and Freedom

The machine—"Not only does it relieve us mechanically of a crushing weight of physical and mental labour; but by the miraculous enhancing of our senses, through its powers of enlargement, penetration and exact measurement, it constantly increases the scope and clarity of our perceptions. It fulfills the dream of all living creatures by satisfying our instinctive craving for the maximum of consciousness with a minimum of effort! Having embarked upon so profitable a path, how can Mankind fail to pursue it?"

Teilhard de Chardin
The Future of Man

Man dreams. Man creates. He changes his environment in search of a better world. He became man when he began to create options beyond those first presented by nature. His hopes for a better tomorrow challenge him to extend his knowledge.

In the U.S. where he has been oriented to the future and open to change, man has extended his basic freedoms. But he continues to hope for a better society. And he suspects that man himself is not a finished product.

A slowdown in the pace of technological innovation should concern us. Yet, most of us are unaware that we are putting less and less effort into innovation and deriving less return.

We might well be expected to understand what technology is and appreciate its material, social, and psychological dimensions. But apathy and antitechnology feelings have become a new morality for some. If we condemn our past and its reliance on innovation, we risk condemning ourselves to a lustreless future.

This is a technological nation that is losing confidence in its ability to manage technology. Rather than constantly redirecting and refocusing it to meet our demands, we tend to get lost in a simplistic debate over whether technology is good or bad.

Technology and society work a complex weave of effects on each other. This society recognizes certain obvious hazards in unbridled technological progress. But it may be vulnerable to the hazards presented by a slowdown or retreat from technological innovation. The odds for turning aspirations into achievements could grow slimmer. One need only take a momentary look backward to see that.

Somehow, ages ago, man learned to control fire. With it, he triumphed over darkness, cold, and a precariously limited diet. He freed himself to wander and, later, to settle in communities. Fire symbolizes both man's continuing process of learning enough of his world to make it more of a home and the freedom to make more of himself.

"In every human soul as it confronts the world there is a sense of the portentous," wrote John Luther Adams.[1] "Our existence, our being here at all, our being in the world, is the really portentous fact for us; it is the sense of the unfamiliar, the strange, the threatening, the sense of not being at home in the world, even when there are no special threats . . . . all technology is the overcoming of the portentousness in things. The bare knowing, the classifying of things by means of laws and general interrelationships represents a repression of their demonic depths, of their incomprehensibility, of their strangeness, their threat to human existence."

Man's brain—his intelligence—permitted him to defend himself against creatures that were stronger, faster, better physically equipped for combat. It enabled him to learn and to adapt to the fierce and changing elements that were hostile to his frail form.

From the beginning, freedom and technology progressed hand in hand. Using the powers of nature to serve his own ends, man freed himself to do more things—some that his ancestors longed to do and some they never dreamed of.

Man's rise to civilization is the story of his untying himself from a direct dependence on nature to increased affluence and freedom. Man has come far from where nature left him. Gestures gave way to speech; then came writing, the telegraph, printing processes, the wireless, television, and communications satellites. The wheel, the sail, airplane wings, and rockets improved the speed and ease of travel.

There are some who say that man is carrying on his own evolution. His style, however, is a departure from nature's ways. Unlike his predecessors, says Lecomte du Nouy, "in order to evolve he must no longer obey nature. He must criticize and control his desires which were previously the only Law."[2]

Once man asked whether an act was good or bad—acquired a conscience—he acquired a new liberty, says Lecomte du Nouy. But this new liberty came with a sense of destiny, a powerful drive to distinguish himself from other animals. This drive has pushed man to know and understand his world. From time to time, he has enslaved his fellow men and suppressed their freedom to inquire, but the drive has never been extinguished.

The brain—with its powers of abstraction—has leaped well beyond the clumsy mechanisms of evolution which preceded it, Lecomte du Nouy maintains. "Thanks to the brain alone, man, in the course of three generations only, has conquered the realm of air, while it took hundreds of thousands of years for animals to achieve the same result through the processes of evolution . . . we see the infinitely small and we see the infinitely remote; we hear the inaudible; we have dwarfed distance and killed physical time."[3]

In recent years, man's conscience has been troubled by the very drive that makes him man. Perhaps he is taking a new turn in his evolution. Rather than viewing himself as a conqueror of nature, he now sees that he can be no more than an equal partner. He realizes that he is still rooted in the material world of animal needs and bound into the interdependent systems of his planet. He longs for the wisdom to guide his use of the knowledge he has acquired. Sensing that he must be more selective in his inquiries and his actions, he hesitates.

There are those who fear that man's technological developments have brought out the worst rather than the best of him. They insist that all technological progress should be stopped. Others personify technology or regard it as an evil force.

"Many people tend to think of technology as being embodied in the machines that we invent and use, but technology is certainly not machines," says Dr. Lewis M. Branscomb, vice-president and chief scientist of IBM Corporation. "Technology is what people do with what they know."[4] Dr. George Bugliarello, president, Polytechnic Institute of New York, sees it as a process—or two processes. "Technology . . . is a social process which generates and combines know-how and people in order to extend the physical range of man. But it is also a biological process because, in enhancing people, it continues to be carried on outside of our bodies and now to a growing extent also inside our bodies. It continues to carry out the process of evolution."[5]

Technology, says James Young, General Electric Company's vice-president of technical resources, "is all the techniques, knowledge, lore, methods, and tools that have helped society survive and improve its life."

Learning The "Why" And The "How"

In his earliest days, man's technology grew through luck, inspiration, and trial and error. Often, he did not understand why his useful innovations worked; he knew only that they worked. Only in the last several centuries has he developed a systematic approach to studying the "whys" of the world around him. He has become scientific—making observations, collecting data and analyzing it, and determining laws and relationships whereby he can predict the outcome of actions.

Science can be thought of as "rules" and technology as "tools," says Dr. Robert Kahn at the University of Michigan's Center for Social Research. We might say that science is the pursuit of knowledge while technology is the use of knowledge. The scientist may pursue knowledge for its own sake but the technologist is utility-oriented.

Technology is not merely the application of science. In fact, it flowered long before science. The wheel and the lever owe nothing to theoretical physics and the bow and arrow were used without knowledge of ballistics, says H. P. Rickman of the University of London.[6] Man knew the "how" before he learned the "why." Increasingly, however, science is preceding technology. By better understanding the rules, we can anticipate ways of putting them to use. Theory sometimes paves the way for practice today. In turn, technology provides the devices scientists need in their pursuit of knowledge.

Both science and technology are the fruits of creativity. While we can appreciate that a scientific advancement may be a creative step forward into new knowledge, we tend to forget that technological innovation may involve all sorts of creative acts. Developing an invention or bringing an innovation into the marketplace demands a blending of knowledge, insights, and anticipation of need. Success may come through careful, systematic figuring or through a flash of insight—a mental picture of the end result. We now see innovation more as a process than a simple act, and creativity may be called for all along the way—in the design of a product, its manufacture, its distribution, its marketing, the manner in which it is used, and in follow-up service.

In many cases today, innovation is not the result of one individual acting as technologist and entrepreneur. It may be the goal for a project carried out by hundreds or even thousands of people working at the many stations along the innovation process.

A Million Hands Weaving

Man's scientific knowledge and technological capabilities have expanded like a giant fabric woven by a million hands working without design. The information needed to support a technological solution to a problem may be developed in a most unlikely place or field of knowledge. Likewise, all sorts of equipment and processes may be called upon to assist the scientist in his studies.

Technological innovation is pulled along primarily by society's demands and, to some extent, pushed by the needs to advance science. Whether it is led by the capabilities pointed out by science or the spoken and unspoken requests of society, we get what we ask for—not what some force dictates. "Throughout history, technology has been the instrument of society," says James Young. "It was spurred by society's needs as society or its representatives interpreted them. Intensive studies show its shortcomings were in the claims society placed on technology, not the power that technology forced upon society."

Obviously, society has not demanded each of the innovations it has received. No one asked Thomas Edison to invent the electric light nor the Wright brothers to build an airplane. There was a perceived demand there, however—perhaps a latent or potential demand. And some inventions have been stumbled upon yet were greeted with great market potential. Alexander Graham Bell developed the telephone as the result of his work with the deaf. Although he was a genius at invention, he admitted that had he known more about electricity he would never have invented the telephone.

Sometimes, society is not ready to accept an innovation. There may be a "cultural lag" until the need is felt, the price is brought low enough, or people are comfortable with it. On the other hand, there are many needs such as in health care where we suffer a technological lag; when the right product comes along the market will be more than ready.

More Return To Society

Frequently, satisfaction, fame, and wealth have come to the individuals who have made technological developments. But individual rewards are infinitesimal compared with the gain shared by all of society. Compare Thomas Edison's fame and fortune with the billions of dollars in benefits bought by millions of people from the companies his inventions spawned. Or compare the millions made by Henry Ford with the trillions of dollars' worth of transportation the world has enjoyed over more than half a century. Like fire, the electric light and the auto have extended man's freedom.

A number of economists have studied innovations and their payoffs to society versus their benefit to the firms or industries executing them. Dr. Edwin Mansfield of the University of Pennsylvania examined seventeen innovations, and he estimates their median rate of return to society has been about 56 percent. This compares with a 25 percent profit before taxes to the industries involved. The higher return to society includes the industry's profit plus price reductions caused by the innovation.

While the numbers may be debatable, the important point is "that the return tends to be higher for the nation as a whole than for the particular industry involved," says Dr. Richard C. Atkinson, director of the National Science Foundation.

Even though the producer using a technology may make profit, "the 'gain' which radiates into the sphere of the consumer is far greater than that of the producer," says Eugen Loebl, once Czechoslovakia's first deputy minister of foreign trade and now a professor of economics and political science in the U.S.[7] He relates the example of a shoe producer who, without applied technology, would have to charge people the equivalent of many hours' work for a pair of shoes. Even at that, there might not be enough shoes produced for everyone who wanted them. "This phenomenon—that 'gain' has the fundamental property of radiating into all pores of the sphere of consumption and of being inherently social—we call the lucroactivity of science," Loebl says.[8]

Yankee Ingenuity

America's high standard of living results from its acceptance of technology. "Yankee ingenuity" is freedom at work. Max Lerner wrote: "The popular word *knowhow* distills the whole American approach, making sure to add the 'how' to the 'know.' That is why science in America is not only a description of how things work but so quickly becomes a technique for making them work."[9]

The U.S. has earned a unique place in history not through its contributions to basic science so much as its exploitation of technology to create a better life. It has shown the ability to innovate on a broad scale. "The American corporation has done its job exceedingly well in transferring technology into products and services and taking them into markets," says Donald Alstadt, president of Lord Corporation. "This has been the wealth-creating process. This process—which has as its byproduct, profit—is magic. For this process to go on, innovation must be present."

From the beginning, with workers in short supply, Americans welcomed new ideas for improving industrial and agricultural output. Technological innovation has been the driving force for making our workers as productive as any in the world. Ed Denison, the noted Brookings Institution authority on economic growth, has determined that from 1929 to 1969 nearly half the increase in national income was due to new techniques, advances in knowledge—in short, technology. The remainder of the increase came as a result of sheer growth in the size of our labor force and in the amount of capital that was invested.[10]

How else could Americans manage to earn such high incomes in relation to the rest of the world, except through multiplying their capabilities with technology? Only Sweden and Norway outrank the U.S. in personal income, and those countries have been more receptive to industrial innovation in recent decades than we have.

When this nation was born, the average income per person was $50 a year. Today, it's more than $6,000 a year. And the workweek has been cut in half! Median family income in the U.S. has nearly doubled since the end of World War II, says the Commerce Department. Measured in 1974 dollars to eliminate the distortion of inflation, median family income went from $6,691 in 1947 to $12,570 in 1975. The rate of improvement for minorities was slightly better than that.[11]

In America, man no longer works long days, every day, throughout a short lifetime. He may spend a third of his years in school, enjoy leisure time during his working years, and look forward to retirement. Survival is no longer dependent on employment. A job provides more than mere economic survival; we look to it for identity and fulfillment.

Twelve-Letter Word

This economic and social prosperity comes from productivity. Productivity is an economics term that may either mean nothing to us or suggest some ruthless scheme for getting more work out of us. It really means obtaining greater output with less input and less waste.

If the boss can get us to work twelve hours a day rather than eight, that's not productivity. That's simply more work. If, however, in an honest day's work we can produce twelve widgets rather than eight, that's productivity improvement. We are being more productive with our time. If the company then has 50 percent more widgets to sell, it makes more profit and can afford to pay us better for that hour's work. With that higher rate of pay, we live more comfortably, work shorter hours, and still earn adequate income. If, on the other hand, we demand higher pay without increasing our productivity, that's inflation. The company has to charge higher prices for its widgets. Wages go up; prices go up. Productivity has gone nowhere and neither has our standard of living.

If we try to create jobs simply by adding people without producing more, we reduce our productivity and our standard of living. French columnist and author Jean-Francois Revel warns that "putting twenty people on a job that one can do only leads to impoverishment. It means that one salary is being split twenty ways, not that twenty jobs are being created, for people can share only what has in fact been produced."[12]

We can increase productivity, of course, by sweating harder at our work. There's quite a difference between a slow ditchdigger and a fast one. But compare the output of the best of them with a backhoe! The big productivity improvements have come, not from working harder, but from investment in machines, the use of brainpower to create better production methods and better designed products, and from training and managing ourselves in more effective ways of getting jobs done.

The nation's output of goods and services climbed 29 percent between 1966 and 1976. The total number of hours worked by the larger workforce, however, increased only 9 percent because the output per manhour had increased. That leverage of greater productivity also helped support a doubling of the total worker compensation package.

The Intangible Benefits Of Technology

"Technology is primarily responsible for a material abundance that allows America to support one-fourth of its population as professional students," points out Lord Corporation's Mr. Alstadt. This is a benefit of productivity that cannot be measured in material terms. In the same way, improvements in productivity have helped free the American woman. As the unfortunate product of centuries of culture, she was burdened with domestic duties. Today, she may complain about the department store computer's mistake on the bill for her wool suit, but it's a suit for which she did not have to card and spin the wool or weave, dye, and sew the fabric. Technology has freed us from the economic need to designate someone to cope with all the chores of running a home.

Woman is finding her place as an equal in society because science is freeing our minds from cultural biases against women. There is no scientific evidence to suggest that women should be ruled from the workplace.

The type and quality of a person's clothing once spoke plainly about his class or financial status. Today, there is little apparent difference between the $89 suit worn by the newest white collar worker and the $350 suit worn by the chairman of the board. The chairman, by the way, may wish his son would wear an $89 suit rather than jeans. And who has the college education—the young man in the suit or the one in the jeans? There's no sure way of telling.

America has never known the plagues that ruined the civilizations of Egypt, Athens, and Rome. Our knowledge of sanitation, nutrition, and medicine plus better living and working conditions doubled our life span in 200 years. We may forget that while 97 percent of our 70,000 housing units have full plumbing today, only 55 percent were so equipped as recently as 1940 (when we had only half as many housing units).

One by one, we have tamed or eliminated many diseases. Gone are diphtheria and smallpox. Childbirth is one-hundredth the risk to mothers that it was in early America. As recently as 25 years ago, families tensed each summer as polio swept through the neighborhood. Science and technology have reduced the number of polio cases in a year to 1/1000 of the level of two decades ago. The number of cases of measles is one-tenth the level of the 1950s. The death rate per 100,000 persons is less today that it was in the 1940s for heart disease, stroke, influenza, pneumonia, and tuberculosis. On the other hand, we manage to create problems for ourselves. The death rate for cirrhosis of the liver and homicide has been on the increase. The incidence of gonorrhea cases reported has tripled in the last twenty years. Drug abuse has become a major killer.

We tend to take technology for granted. When we do think of it, we often think in terms of products—cars, television sets, microwave ovens. We can condemn it by listing what we may regard as useless or diversionary products such as electric toothbrushes, electric backscratchers, and battery-operated toys. We'd like to do without some of the products of technology. There are others we would like to keep only if we could eliminate their adverse side effects or their social impact. Whether they resulted from market demand or government-sponsored research and development, there are some technological innovations that some, if not all, of us regret.

There are many reasons, as we shall discuss in the following chapters, why we will have to be more selective in our use of science and technology. As always, man must go on innovating, but he cannot trust that all innovation is good and that everything will turn out all right.

At the same time, we cannot take for granted the benefits of technology while we are deliberately or unknowingly placing limitations on our ability to innovate. Neither blind acceptance nor aversion to technology will prepare us for the complex questions we have to resolve.

We have reached a level of affluence at which we can afford to direct more attention to creating the good life. Now that we have made technology so much a part of our lives, we have to ask how we will change our lives and our technology for the better.

Our Revolutionary Society

It is not change that we must avoid but undesirable change—or stagnation. The key to America's greatness is not in the level it has attained but in its quest for greatness. We are a changing, evolving, perhaps revolutionary society. Our technology has presented us with options and it is up to us to make the best from them.

Revel sees the U.S. as the one nation in which an endless revolution is taking place. The reason the U.S. is changing and improving, he writes, is that freedom and technology have nurtured one another here.[13] Two decades ago, Max Lerner wrote: "More than anything else, this pace of technological change is what gives America its revolutionary character today."[14]

There is a close relationship between the principles of science and those of a free society. The values of free inquiry, free thought, free speech, and tolerance seem self-evident, says Jacob Bronowski. "But they are self-evident, that is, they are logical needs, only where men are committed to explore the truth: in a scientific society."[15]

Freedom would be a hollow term without the time, health, education, transportation, and communications we have derived from our technology. Dr. Edward E. David Jr., president of Exxon Research and Engineering Company and a former science adviser to the President reminds us that technology "reforms our lives. The most diverse life-styles of anyplace in the world—that's a product of technology."

2 Less Innovation, Limited Options

"Societies will, of course, wish to exercise prudence in deciding which technologies—that is, which applications of science—are to be pursued and which not. But without funding basic research, without supporting the acquisition of knowledge for its own sake, our options become dangerously limited."

Carl Sagan
The Dragons of Eden: Speculations on the Evolution of Human Intelligence

"The era when we were a strong technological giant is over," says Paul F. Chenea, vice-president in charge of General Motors Corporation's research laboratories. "We don't dominate the world anymore technologically."

Taking the world view, that is good and is to be expected as other nations advance. "If other countries are going to bring their standards of living up, that will happen," says Dr. Chenea. But the U.S. will find the already tough world competition getting even tougher. Countries with cheap labor and borrowed technology will be extremely rough to compete with.

The U.S. still holds the world lead in technology, but we are slowing in our rate of innovation, and the gap is closing. The tides of trade could shift dramatically against the U.S. in such things as autos, electronics, and textiles. "We'll have to learn to compete. Our lifestyles are going to have to be modified. We'll have to return to increases in productivity and pride in our work," says Dr. Chenea.

Arthur Kantrowitz, president of Avco-Everett Research Laboratory Inc., is convinced that "there is a clear and present danger from the deceleration of American technology. There can be little doubt that . . .the resources that we have developed in new technology and the optimism which wants growth have declined in recent years."[1]

Many leaders in industry, science, and government are deeply concerned that we are losing momentum in our technological effort. We are doing too many things that could kill our innovative spirit and ability.

Even President Carter has shared that concern. "The quality of scientific equipment has been falling off rapidly in recent years," he said in November 1977 at the National Science Medal award ceremonies. Troubled by the decline in the number of research centers, he said, "We want to make sure that the climate for research and development in our country is enhanced."

The President directed federal agencies to boost their spending for research and development and launched a study of the reasons for the slowdown in the nation's research and development (R&D) effort. His science adviser, Dr. Frank Press, says: "If industry is not innovating as it has historically or compared to other nations, we want to know the reasons why. Is it a question of government disincentives? Is it a question of tax policy or antitrust legislation? Is it a matter of regulations and the need for industry to invest in defensive research to protect their existing products against government regulation? We also want to know what foreign governments are doing to spur innovation in their countries."

In his fiscal 1979 budget message, the President said: "The administration believes that the continued advancement of basic knowledge in all fields of science is essential to the continued growth of the economy and to the understanding and ultimate solutions of problems in many areas of national concern, such as health, energy, environment, and national defense."

The budget contained a 6 percent increase for research and development. Considering the effects of inflation, however, that hardly looked like an increase. Furthermore, two thirds of the added dollars were designated for defense R&D.

Of the $27.9 billion in obligations for R&D in fiscal 1979, $10.4 billion were for civilian (nondefense and nonspace) purposes—an increase of only 3 percent. The concern expressed by administration officials had not yet been translated into dollars for research through government channels or any action that would serve as an incentive to private research.

In January of 1978, a group of distinguished scientists, many of whom were Nobel laureates, testified before a Senate subcommittee and expressed their concern over the lack of consistent science and health policy in the U.S. and the inadequacy of funding for basic research. They pointed out that funds spent for basic research had dropped despite rising costs, industry employed 43,000 fewer scientists in 1976 than it did in

1970, and the percentage of the gross national product for research had dropped substantially since 1967 while it was rising fast in Japan, West Germany, and the USSR.

Seed Money

The seed money for technological innovation—research and development spending—has been shrunk by inflation. In 1976, American industry and government spent a total of $38 billion for these purposes. The 74 percent gain from 1966 to 1976 figures out to zero after inflation is considered. R&D funding, therefore, constitutes a shrinking share of our growing economy. It fell from 2.9 percent of our gross national product in 1966 to under 2.2 percent in 1976. During that same time, Japan's percentage rose to 2 percent from 1.5 percent. This may explain, in large part, why Japanese products are more and more attractive in terms of innovativeness and price.

More than half of our R&D funds come from government. Although government funding has increased steadily since 1971, these dollars represent a decreasing share of the federal budget. In the mid-sixties, nearly 13 cents out of every dollar spent by the federal government went for R&D. By the mid-seventies, this share had dropped to 6 cents. Big cutbacks in defense and space programs account for the drop.

The R&D financed by industry has barely held even with inflation in recent years. Measured in constant dollars, it flattens out a little below the peak it reached in the sixties.

Looking For The Quick Payoff

Research and development includes a range of efforts from "basic research" where new knowledge is sought, to "applied research" which works toward practical application of knowledge, and "development" where new products and services take their final shape.

Basic research has suffered most in the R&D slump. At $4.8 billion in 1976, its 68 percent gain in a decade converts to a 4 percent drop when inflation is taken into account. Industry's share of the spending for basic research lost considerable ground; it increased its allocations 44 percent, allowing a 17 percent slippage after inflation.

"Since risk increases as one approaches the basic research end of the

R&D spectrum, it is not hard to understand industry's preference for R&D with shorter term payoffs. In fact, business appears to have shortened the time horizon for R&D payoffs to as little as two to three years in contrast to the 10 to 15 years required for basic research," says Dr. Richard Atkinson of the National Science Foundation.

Business has traditionally been committed to the shorter term R&D, leaving basic research largely up to the government and universities. A relatively few firms have the inclination or the money to invest in long range research in their own labs or those of a university or other research institution. But in recent years, there has been increasing interest in the short term—a shorter short term than usual.

Under pressure to maintain profits in the face of rising costs and new constraints, industry has taken a hard look at R&D spending and demanded measurable results. Longer range research, hard to measure and obviously not a producer of immediate results, has been getting less and less attention. About half of industry R&D money, surveys show, is being devoted to improving existing products while a third is invested in developing new products.

A basic research dollar is invested with extreme care not only because it may be a dollar that doesn't produce any results the company can use but because, if successful, it may involve the spending of ten more dollars for development and then a hundred more to gear up for producing and marketing the new product.

A researcher with one large company recalls how his firm was a strong advocate of basic or, at least, fairly fundamental research in the fifties. But when the profit pinch came in the sixties, the firm began emphasizing short-term developmental work. A manager in a large corporate laboratory says he has noticed that divisions sponsoring research in his lab will no longer fund any work that can't promise a payoff within one or two years.

Government-funded R&D has also focused on the short term. "Many in the private sector have complained about (legislation) which requires that funds provided by the Defense Department to companies for independent, long-term R&D must be spent on mission-oriented work," the Commerce Department reported in a 1977 draft study on technology policy.[2]

This "doctrine of relevance" has hit present and future innovation hard, says Dr. Jacob Goldman, group vice-president and chief scientist at

Xerox Corporation. He reminds us that "it was the military that made possible transistors, microprocessors, and computers. The electronics revolution owes a lot to work supported by the military, even when the relevance was not so great."

Defensive R&D

Another shift that has cut into the limited dollars available for innovation is a defensive move that companies have made to comply with regulations on such things as pollution control, worker safety, and product reliability. It amounts to more than 10 percent of the R&D effort for Union Carbide Corporation, estimates Dr. Thomas R. Miller, vice-president. An additional 10 to 20 percent has been directed to energy-related problems, he adds. But this isn't all "cost," he points out. "you can get some good savings out of that."

In a letter to the President, Dr. R. E. Boni, Armco Steel Corporation's vice-president for research and technology, pointed out in early 1977 that the amount of R&D money "spent in response to federal regulations is increasing at an alarmingly rapid rate." Sampling 71 major companies, he found that "reactive research to various legislative requirements is growing at an annual rate of increase of 10 percent to 17 percent. Energy-related spending was climbing at a median rate of 26 percent. "These statistics mean that R&D devoted to growth and diversification of U.S. industry are decreasing at corresponding rates," he wrote the President.

Industry managers do not begrudge R&D spending for meaningful, reasonable environmental and safety objectives. They are people first, managers second, and they treasure their health as much as anyone. But they do warn that many of the costs are unreasonable. Even worse, because of uncertainty regarding possible new regulations or legislation at any time, companies are forced into defensive positions, devoting more R&D dollars to staying out of trouble and fewer to "getting into" new products.

While the goals of these regulations are worth pursuing and response to them is generating some innovation, we should be aware that they are diverting money from the traditional forms of innovation that gave us new products, new industries, and new jobs. We must realize that we are unknowingly making a trade-off—that we are paying for environmental and safety considerations in the products we buy today and in terms of the new products and new jobs we may not have tomorrow.

Of course, some new products have been born as a result of the need to attend to environmental and safety considerations. But we have to question whether we have gotten out money's worth. Unless we can be sure that the cost of compliance with regulations is efficiently translated into the results we want, we may be wasting money that could be meeting other vital needs.

Industry is now channeling one-tenth of the money it invests in plant and equipment into pollution control devices and systems. This places a strain on the ability to expand and to improve efficiency. It also leaves fewer dollars for R&D.

Environmental protection and human safety are high priority objectives. But they are not the only ones we have to consider. If, in serving them, we lose our innovative spirit and strength, we lessen our chances of meeting any of our objectives. "Basic research . . . provides much of the fundamental knowledge upon which modern scientifically based industry is built," says Dr. Thomas Vanderslice, vice-president and group executive of General Electric Company. "Much of what we expect science and technology to provide in the future will be slower in coming unless present trends are reversed."

Less For Productivity

The slowdown in R&D could affect our productivity, thereby threatening our ability to improve our standard of living and to compete internationally.

"In the 1950s and early 1960s—the period of greatest jumps in productivity and the hey-day of U.S. innovation—the environment for innovation was nearly ideal," says Dr. Glenn Brown, vice-president of research and engineering for Sohio. "Most major universities had active, vital scientific research programs which drew the top secondary school students. Industry was searching for opportunities to invest in new and growing areas. The consumer was eager to receive each new innovation—since most of them gave him an immediate improvement in comfort or finances. The Government, which has the responsibility to protect the consumer, underplayed, if anything, its role as a regulator."

Since that heyday, productivity in the U.S. has not been improving at as fast a rate. Dr. Brown notes that it improved an average of 3.9 percent a year between 1961 and 1965. During the next five years, the rate dropped to 2 percent. And in the 1971 to 1975 period, it fell off to 1.6 percent. Even the most optimistic forecasts for productivity improvement in the next several years are on the order of 2.5 percent—not what it used to be and not what it should be to offset rising wages.

This coincides with a period of weakened R&D spending and a diversion of funds from those potential innovations that could improve productivity. By contrast, Japan and Germany devote a larger proportion of their R&D effort to civilian uses—particularly those that can lead to giant steps forward in products or manufacturing processes.

The trend in R&D spending looks even more troublesome if one considers that a great deal of today's research involves gigantic outlays. We have reached barriers that are harder to penetrate in science, says Robert Colodny, scientist and teacher of the history of science at the University of Pittsburgh. Gone are the days of the tinkerer who could make important inventions, often without understanding why they worked. Today, it takes teams of highly educated specialists to make advances which sometimes only they can understand.

"Science has been industrialized," says Dr. Colodny. Today, we have to mobilize large amounts of capital and build expensive research facilities. The production and marketing of a new idea may take billions of dollars. In simpler days, a scientist could work with a blackboard and chalk plus some equipment he might build himself.

The National Science Foundation says the largest share of major innovations comes from large companies—those with 10,000 or more employees. NSF also notes that the number of innovations from large companies has been increasing over the years, in both absolute and relative terms.[3]

Where Are The Big Ideas?

To gauge the slowdown in American technological innovation, there are two sides to look at, says Dr. Goldman at Xerox. You can look at the dollars invested, and they are certainly on the decline. You can also look at the output—the innovations that materialize.

"No matter which way you try slicing it, I think you end up finding that there has been a decrease in the rate of innovation," he is convinced. Instead, he sees a "shift to simply improving the old." Dr. Brown at Sohio agrees. It has been well publicized that "numerous U.S. companies are foregoing the search for major innovations . . . for incremental improvement in existing products and processes which hold out greater potential for short range return."

Dr. Goldman suggests taking a look back to the fifties and sixties. "You come up with such very striking and staggering inventions as xerography (he could hardly omit it), the Polaroid camera, the microprocessor, the whole electronics revolution, color television, and some of the very revolutionary new drugs. Try to make such a list for the last five or six years," he challenges.

Of the innovations we have seen, an increasing flow has been coming from abroad. The U.S. Patent Office issued slightly fewer patents to U.S. individuals and firms in 1973 than it did in 1963, but patents issued to foreign nationals more than doubled during those years.

In a special study done for the National Science Foundation, Gellman Research Associates found the U.S. is responsible for a declining share of the world's major new products and processes. Of 500 major innovations brought to market during 1953–1973, the U.S. accounted for 82 percent in the late fifties. We were the technological giant. But by the mid-sixties our share had slipped to 55 percent.[4]

U.S. industry has traditionally been a borrower of fundamental ideas. Many of our basic production processes were conceived in Europe, for example. But we generally have managed to capitalize on them by adding good developmental work and the managerial expertise to get them on stream. We have also been helped by the fact that the U.S. offers a large, ready market for a new product or the output of a new process.

The U.S. steel industry is generally regarded by businessmen as relatively conservative in its approach to R&D. Most of the major innovations in iron and steelmaking over the last quarter century first went into commercial application in other countries and then were adopted here. A study by the American Iron and Steel Institute shows that "information on new technology diffuses throughout the world very rapidly." The flow, however, seems to be almost entirely to the U.S. The basic oxygen process for making steel was first used commercially in Austria in 1952; it appeared in the U.S. in 1954. Hydrochloric pickling to clean hot rolled bands of steel appeared here in 1963—three years after it went commercial in Germany. Japan initiated the use of giant blast furnaces for reducing ores in 1967; Bethlehem Steel followed two years later. There was a four year lag between Japan's use of fully continuous cold rolling mills and National Steel's use of them here.

"Being the first to implement commercially a developing technology is not necessarily the best economic decision," states the report. "In most cases, technology development is evolutionary, and major significant advances are sometimes made painfully and at high cost during the operation of production equipment."

American steelmakers have long felt at a disadvantage in taking risks with new technology. Raising capital is more difficult here than in some nations—particularly Japan where the financial sector and government work closely with industry to achieve national economic objectives.

It is also easier to incorporate new technology when adding to capacity which is what other nations did while U.S. steel companies did not need significant additional capacity until the last few years. The conservatism of this mature industry coupled with a less than ideal climate for risk-taking has rendered the U.S. steel industry a follower rather than a leader.

The uptrend in patents registered in the U.S. by foreign inventors reflects not only the strength of foreign science but the general receptiveness of U.S. firms to buy their know-how wherever they can get it because of their reluctance to gamble and the intensified need for quick results.

J. Herbert Hollomon, director of the Center for Policy Alternatives, is not concerned about a drop in R&D. "What R&D does is contribute to the capability of mankind, provide the advancing tools of technology, but in and of itself does nothing. People are concerned in the country, and I think rightly so, about the level of basic research, but . . . the world's basic research is available to everybody in the world for a first approximation and usually quickly."[5]

Marching a half step behind other nations in basic innovations and then leapfrogging with good development work, the U.S. might hope to be competitive in the world marketplace. But if our receptiveness or ability to implement innovations should be impaired, we will be in trouble.

We often hear that the time between invention and implementation of an innovation is growing shorter and shorter—that we are racing to get new ideas on the market (regardless of where they come from). The National Science Foundation notes, however, that the invention-innovation interval did shorten from the early fifties to the early sixties. But it stretched out from the mid-sixties to the early seventies.[6] Furthermore, NSF observes, the interval seems to correlate with an industry's R&D intensity. If this is true, a slowdown in the nation's overall R&D activity could signal a slowdown in our whole innovation process.

NSF data also show an apparent correlation between a nation's economic health and the invention-innovation interval. The nation with the shortest interval: Japan, at an average of 3.6 years, then West Germany at 5.6 years, the U.S. with 6.4 years, and the United Kingdom, 7.5 years.[7]

Is The Foundation Eroding?

Dr. Colodny, who has studied scientific development throughout the world, is concerned that the underlying foundation of science is inadequate to support advancement of our technological society. We lack conceptual innovation, he feels. The last big theories—relativity and quantum mechanics—date back to 1905 and 1913. The science for such things as space vehicles, computers, and lasers was developed years ago, he points out. "The glory of this era is simply the application of earlier knowledge."

He is particularly concerned that not enough of the brightest people are entering the fields of science. A shrinking supply of bright new people to put something in the "bank" threatens the long-term prospects for innovation. The number of people enrolled in programs for advanced university degrees in science and engineering peaked in the late sixties and has tapered off by about 10 percent since then. In 1975, 3 percent of the people enrolled for advanced degrees were in the physical or environmental sciences compared with 9 percent a decade earlier.

Industry goes through periods of crying for engineers and then sometimes finds that it has a temporary excess. But it may do a great deal of crying in the coming years; 6 percent of 1975's advanced degree candidates were in engineering compared with 11 percent in 1965.

Science and technology go through cycles of favor and disfavor among students. But there may be a longer-term depressant at work in addition to the normal attitudinal shift. The high cost of education poses a strong deterrent to the person who would devote several years to study beyond the bachelor's degree. This, coupled with a decline in the research funds granted to universities, results not only in less research being done but less opportunity to train people in research. The postdoctoral work which traditionally provides much of the muscle in basic research has been hit especially hard by tightened industry and government pursestrings. Faculty and graduate assistants are swayed, too, by the need to work on projects that promise quick payoffs in order to obtain funding.

Are We Topping Out?

"We have seen a topping out of technology," believes Dr. James L. Heskett, professor of business, Harvard Business School. "We won't have the technological advances we've had in the last twenty or thirty years."

But there are some who disagree. "We haven't begun to see the impact of technological innovation," argues Dr. D. Bruce Merrifield, vice-president for technology, Continental Group Inc. "We'll see the synthesis of many ideas that will impact things in ways nobody expected."

In terms of the application of current knowledge to new products or modifications, Dr. Merrifield may be right. For example, Dr. Robert Noyce, chairman of Intel Corporation, says: "Electronics are going to be everywhere in the next decade. Annual usage of electronic functions should go up by a factor of 100 and prices will go down by a factor of 20." He points to mechanical functions being replaced by electronics in watches, automobile engine controls, calculators, thermostats for kitchen ranges, burglar alarm systems, games on the TV set, the dial on the telephone. "Electronic mail is certainly going to come," he says.

Leaders in the electronics industry are finding more and more applications for the knowledge and basic hardware they now possess. Their companies may flood us with change in the next few years. Microprocessors, following on the heels of computers, are causing major changes in manufacturing and in some of the service industries like banking. At the consumer level, the changes will be even more apparent in products such as those listed by Dr. Noyce.

This flow of new "gadgets"—some with important value, others of questionable value—might act as a smokescreen. If we continue to borrow from the "bank" without putting anything into it, the apparent pace of innovation will be strong until we run through our account.

Too Little In The Bank

"What confuses us is the period from 1930 to 1950," says Robert M. Coquillette, executive vice-president, W. R. Grace & Company. "That was the period in which we developed a large backlog of innovation which we weren't able to commercialize because of the depression and which was built up by the war impetus. There wasn't time for peacetime applications in chemistry and electronics. Then these things appeared from 1950 to 1970."

We might add the impetus of the space effort which came on the heels of World War II and the cold war. This kept alive both the spirit of innovation and provided real dollars for research and development. Although the innovation effort was devoted to military and space work, there were sufficient spinoffs to support the civilian markets with new ideas.

"There hasn't been an interruption to the process, but we have run through the backlog," says Mr. Coquillette. "We have no bank of ideas." While he looks back to the bulge in ideas and sees today's level of activity as a more normal rate of innovation, Dr. Merrifield sees the present pace as a momentary drop below normal. Ecological and energy considerations are causing a diversion that could last another five to ten years, Merrifield believes.

Whether the slowdown in innovation is a back-to-normal or a below-normal state, the rate of idea propagation will not sustain the economic growth and productivity improvement we enjoyed in the fifties and sixties. It could bring an economic malaise that is hard to diagnose, harder to explain to the public. It could thwart our efforts to create more jobs and control inflation.

No single indicator—patents, R&D spending, the U.S. share of major innovation, the number of students enrolled in science and engineering—should be read as a harbinger of doom for the entire innovation effort. But "where there's smoke, there's fire." With indicator after indicator turning down and the attitudinal climate posing a threat to innovation, we may be near the point where there is smoke but no fire.

Despite his concern over the indicators and the atmosphere in which he guides his business, Dr. Vanderslice at General Electric looks positively to the future. "I prefer to think we are going through a transition—a transition marked to some extent by the dismantling of a scientific and technological infrastructure that has served us well in two world wars, and in the cold war that followed, and a rebuilding to serve our needs in the last decades of this century and the first of the next."

We cannot afford a battle between those who would recast technology as it was and those who want less or no technology at all. We can afford neither to let technology fly recklessly on the winds of chance nor to lock ourselves in a rigidity that would leave us vulnerable to the winds of change. We need the resiliency to make a transition—to refocus our technology, selecting from the options it gives us for economic and social progress.

3 Growing List of Wants

"Be fruitful and multiply, and fill the earth and subdue it; and have dominion over the fish of the sea and over the birds of the air and over every living thing that moves upon the earth."

Genesis 1:28

In a society characterized by change and the search for improvement, a slowdown in technological innovation should prompt us to examine the full range of causes and consequences. We cannot dismiss it as a matter of concern for technologists or businessmen; it raises questions about the nature of our society and its wants. Even in this "land of plenty," not all our wants have been satisfied.

Americans concerned about being overweight greatly outnumber those facing starvation. Because of the increased productivity of our farms it takes only 2 percent of our population to operate them and feed the rest of us. Because of the increased productivity throughout our economy, Americans have been spending a smaller and smaller portion of their incomes for food. We're a well-fed nation.

Farm mechanization and improved hybrids of animals and plants have played a big part in increased food production and reducing prices from what they might have been. Insecticides and herbicides protect us from the loss of 30 to 50 percent of major food crops, says John W. Hanley, chairman and president of Monsanto Company. Chemical additives stop rotting, disease, and loss of nutritional value in processing, shipment, and storage.

We haven't totally forgotten that technology has made food plentiful and relatively inexpensive. When we think of the possibilities of population growth, we expect that technology will somehow supply our needs. As our concern over junk food and food additives grows, we demand that technology not only keep us well fed but properly nourished.

For billions of people in other lands, food is a daily life and death concern. Half a billion children are suffering from malnutrition or dying of starvation.

There is an imbalance of food in the world because some nations have applied enough technology to produce as much or more food than they consume. The well-fed nations have a high level of technology which permits them to maximize their domestic food production and to produce wealth in other forms which can be traded for food. It is no coincidence that in countries where food is short, technology is primitive. Even the technologically backward nation blessed with rich natural resources to sell is in a poor position to feed all its people well; wealth tends to be concentrated in the hands of a few and, without a balanced economy for distributing goods and income, many people go hungry.

Whether we do it for altruistic reasons or for self-protection against have-not nations who might retaliate, we will have to attend to the world food problem. But neither good intentions nor fear will do the job. It will take technology—not only more industralization of food production and processing but totally new technologies.

"Poor people can't help poor people. Poor nations cannot help poor nations," says Richard de Vos, president of Amway Corporation. The people who produce have to do the "human thing"—share what they produce, he says. But we are talking about sharing wealth—not poverty. That means improving our technology to provide better and better ways to raise the productivity of our lands, perhaps even raise food not dependent on the land. We must share not only food but the know-how for self-sufficiency.

Science Versus Starvation

Science and technological innovation worked food production miracles in the last century, but we cannot expect our present state of development to cope with the problems that exist today, much less those that lie a few years ahead. Battelle Memorial Institute, a large and highly respected research organization, warns: "Beyond 1985, there is little prospect that further productivity improvements will be possible unless there is more done to add to the store of fundamental scientific knowledge that has been the basis for the improvement."[1]

Battelle's researchers are at work in a number of areas that relate to the food problem. They are seeking improvements in irrigation and methods of preventing loss of soil fertility. They are concerned, too, with the ecological dangers posed by pesticides and fertilizers. Energy scarcity poses new problems that will have an impact on the supply of food, they say. For that reason, they are working on such things as: methods to produce higher yields of sugarcane and sweet sorghum which can be used as feedstocks for energy production, innovations in wet-corn milling to reduce energy consumption in the process, development of a 50-horsepower irrigation pump powered by sunlight, and use of water hyacinth plants as an energy source.

S. H. Wittwer, director of the Michigan State University Agricultural

Experiment Station in East Lansing, observes that the rapid increases in farm yields since World War II have come through technologies dependent on fossil energy. But these inputs—fertilizers, pesticides, irrigation, mechanization, and new seeds—"are becoming increasingly costly, subject to more constraints, and less available."[2]

He says there could be technologies available in a few years to provide high-level food production without pollution, require less capital investment, and restore rather than deplete the earth's resources. This, however, would have to come through basic research into the biological processes that limit the productivity of crops and livestock.

One may believe the pessimistic forecasts of a world population explosion, taking us from the present 4 billion persons to 10 billion or 20 billion in the foreseeable future. Or one may be more optimistic and foresee a tapering off in the growth rate by the early twenty-first century. In either case, we see the certainty of billions of starving people unless something is done to boost the food supply dramatically.

The experience of the technologically and economically advanced nations suggests that population modulation will come with affluence. The birth control practiced in the affluent nations is itself a form of technology. Whatever the means used, it reflects man's knowledge of what to do—or what not to do. Technology offers some hope that the demand for food may not continue its runaway growth. But it offers more hope on the supply side by directly attending to the food problem. It could be brought to bear, for example, on the insects, rodents, and disease that destroy as much as half of the world's food production.

If we expect the world to be fed, we cannot turn our backs on scientific research and technological innovation. Likewise, we expect better and better health care. We would not want to give up the technologies that bring us eyeglasses or dialysis machines. To the contrary, we hope innovations lie ahead which will bring us better devices, or even eliminate the need for devices, to improve the vision of the majority of Americans and keep 30,000 kidney patients alive.

We have come a long way from the peg leg to biomechanical replacements for parts of the body damaged by disease or accident. The cardiac pacemaker, miniaturized diagnostic equipment, and voice-controlled wheelchairs have extended the horizon of our hopes for triumphs over physical limitations.

Not only do we hope for better health care, we expect its costs to be controlled so it will be available to all. Another billion dollars to double our medical research might dramatically reduce the nearly $200 billion we spend each year for medical care as well as vastly improve its effectiveness.

The Only Ethical Course

On many fronts, our expectations set in motion the demand for more and better technology. Without technological innovations of the right kind, how will we:

- Upgrade the safety and reliability of our products?
- Combat the pollution we create in our industries and in our daily living?
- Reduce disease and hunger?
- Rebuild our cities?
- Compete successfully in a world market?
- Minimize our demands on the earth's resources?

"More technology, not less, is the only ethical course," believes General Electric's James Young. "More to feed, clothe, house, and provide for the enjoyment, medical care, and service needs of an expanding maturing population. More to help society fulfill the aspirations of the less privileged. More to limit and to reverse the past effect of technology upon the environment. More to conserve resources and to recycle wastes."

A History Of Creating Resources

Implementation of many of our technologies requires the input of natural resources. We, therefore, think of technology as a consumer of resources. In our resource-rich country, technology was first called upon to maximize the output of labor. It multiplied man's muscle and brainpower. Later, it helped reduce hazardous or degrading work. Now it is being called upon to play a new role: maximizing the use of our resources. Actually, this is not a new role. It is a role that is gaining prominence but one that technology has always filled.

A great deal of technology's contribution to man's welfare has been in the development of new materials. Who would have guessed a century ago that aluminum would become a common material for inexpensive pots and pans or a wrapper for leftover foods? Who would have guessed that the resource that would upset world economics in the 1970s would be petroleum—a liquid that had little use in the mid-nineteenth century other than for some questionable medicinal purposes?

In the 1930s, rubber would have been high on the list of critical materials needed to fight a war. But when Japan cut off Southeast Asian sources of supply, our technology permitted us to develop synthetic rubber that was superior to the real thing. An item that would not have been on that list—plastics—has now become a staple. This new family of materials has moved into applications once filled by metals, wood, or paper and even opened the doors for creation of products never seen before.

We tend to think of resources in terms of the technology and economics of the present. "Natural resources . . . cannot be catalogued in geographic or geological terms alone," says Nathan Rosenberg, an economics professor at Stanford University who takes his fellow economists to task for failing to recognize "how profoundly" technological changes continually redefine what resources are.[3] "We don't know what the important resources will be 50 years from now," says C. Lester Hogan, vice chairman of Fairchild Camera & Instrument Corporation.

Our future material requirements will partly be met by applying technology to extend the supply of today's materials. As technology improves and economics change, low grade ores or hard-to-get petroleum will become resources that we formerly ignored. "Contrary to what many believe, technology has not hastened the depletion of our resources," says Fletcher L. Byrom, chairman of Koppers Company Inc. "It has made it possible for us to get eight times more energy from a ton of coal than we did in 1900."

We will also attempt to meet our requirements through technologies which may develop alternative materials or new applications for those which we now have.

Even the conservation of our resources calls for more innovation. In the U.S., perhaps more than anywhere else, materials generally have gone through one round of usage. We flattened tin cans and rolled balls of used foil during World War II, but reusing materials was something to be put up with "just for the duration." Once again, however, economic conditions are favoring, if not forcing, the conservation of our resources. As supplies of conventional materials grow tighter, prices rise, and techniques for scrap recovery improve, we find ourselves getting heavily committed to recycling.

We are already significantly dependent on recycling for some of our key materials. Nearly half of our copper comes from scrap, as does half of our lead, one-fifth to one-fourth of our aluminum, and one-fifth or more of our paper—enough to spare 200 million trees each year.

The energy saved in reprocessing scrap compared with production from virgin materials is substantial and, in itself, will change the economic odds more and more in favor of reusing our throwaways. Government studies show that more than 60 percent of the energy needed to produce iron and steel or paper can be saved by using recycled material. Even more astonishing are the numbers for copper—87 percent savings in energy—or aluminum, 96 percent.

Up to now, the scrap that has been recycled has chiefly been the large lots of paper, steel, or other material that could be readily collected and sorted. But we are now turning to some of the harder-to-get-at sources of scrap. Plain old garbage is now being segregated into valuable resources. Since 1970 resource recovery from municipal waste has jumped from the concept stage through adolescence into initial years of production, says Edwin D. Dodd, chairman and chief executive officer of Owens-Illinois Inc. "The payoff will come in the decade of the 1980s and beyond," he believes.

By 1975, four or five prototype municipal resource recovery systems were on line, he says. By 1977 there were about 15 in operation and 20 more planned or started. By the end of the decade, he estimates, resource recovery plants will be processing more than 6 million tons of refuse a year and plants for processing another 3 million tons will be in the design or construction stage.

More With Less

Product design and architecture are beginning to reflect efforts to conserve energy. The automobile industry is in the process of meeting our transportation needs within the constraints of lower gasoline consumption. As a "by-product," its demands are lessened for some materials—sheet steel for bodies—and increased for others—more lightweight components of aluminum or plastic.

Homes, offices, and factories are being designed to provide greater insulation, take advantage of solar energy, and balance heating and lighting requirements more efficiently. With technological innovation being sought to reduce energy requirements, the resultant demand for some materials—such as insulation—goes up while it declines for others.

In general, the earth's resources are supporting more and more people in better style and answering to more of their wants. The enriched lives of a large share of humanity, despite a decline in material resources per person, is due to "ephemeralization," Buckminster Fuller has written.[4]

"Ephemeralization" is his term for doing more with less on a dramatic scale. One of his best known examples is the considerably lesser material requirements for a communications satellite than for the many trans-oceanic cables it can replace.

Fiber optics, the use of glass fibers to carry pulses of light, is an emerging example of ephemeralization. They can carry far more messages than a wire of comparable diameter. A more sweeping example: the use of communications systems to reduce our needs to be transported about and thereby possibly reduce the requirements for materials and energy.

These are ways that technology gives us new options or opportunities to meet needs. We can, if we choose, replace a product or system with something that requires less material input.

"There is almost no limit to how much of our economy can be made up of high-technology products," says Richard B. Tullis, referring to material resources. Tullis, chairman of Harris Corporation, has seen his own company grow from the production of printing equipment to a broad scope of electronics and communication equipment. "High technology is packaged thought," he says. "The printing press was primarily labor and materials; the computer is thought and a little labor and materials; the integrated-circuit chip is thought and very little labor and materials."

"High technology," by the way, does not always appear in the form of computers or jet aircraft or integrated circuits. Dr. Guyford Stever, former director of the National Science Foundation, points out that "wheat is a high technology product." We don't brew up shiploads of wheat in the laboratory, but many technological innovations have gone into improving agricultural products and their yields.

Use The "Best Available Technology"

Concern for the environment is calling upon technology to serve still another end. A relatively new term—"best available technology"—has been minted by regulatory agencies in their stipulations as to what industry must implement in the quest for a cleaner environment. The same notion applies to product safety and to health and safety in the workplace. An industrial company today must use not only all *its* know-how but all the know-how extant to meet the objectives of safety and cleanliness.

In the scramble to use technology to meet all these objectives, we run the danger of coming up with quick-fix solutions. Technology can give us a quick answer here and a quick answer there. But, in recent years, we have learned a valuable lesson that shouldn't be overlooked simply because needs for improvement are urgent; there are long-range, global consequences for many of the things we do. As we extend technological innovation into the areas society wants treated, we first need a greater understanding of the big picture to ensure we don't make mistakes. Technology can give us the tools, but we need to learn more about the rules. We need more scientific research to help us use our technology more intelligently. Technological innovation, per se, is neither good nor bad. It is up to us to make wise choices in the form and degree of the technology we use.

Long List Of Wants

Whatever our economic and social goals—jobs, control of inflation, more leisure time—we will need more technological innovation to bring us closer to them. Like it or not, our lives are shaped by economic realities and technology has become an increasing factor in economics. We have a long list of problems and wants. Technology can't deliver solutions for all of them but it can, if used properly, improve the economic basis and thereby let us address a higher order of problems.

Although the U.S. enjoys a high standard of living, it is neither the ultimate nor even the best on many counts. It ranks seventh in life expectancy and about fifteenth in infant survival. It is outranked by nearly all the other industrialized nations in the number of hospital beds per 1,000 persons. Our streets are not the safest in the world. Our education systems cry for improvement.

The list of needs and wants never seems to get any shorter. We continually need more know-how to cope with them.

How do we raise our standard of living? Only increases in productivity will do that, says Donald Alstadt. "When I see the productivity of a country going up, I see the standard of living going up. When productivity is flat, the standard of living is to be, at best, flat."

Improvement in our standard of living—whether measured in terms of education, health care, or whatever—necessitates more being done for each of us. And that means more work being done by the sweat of our brow or through our technology.

Productivity improvement through technology can help us create jobs for a growing workforce in a world economy where workers compete by being effective. While we can use our wealth-creating ability to help the less fortunate in the world, we have to be good enough at it to maintain a balance against other wealth-creating nations. We can't share wealth if we don't produce it, and we can't produce wealth unless we are competitive.

World Competition For Jobs

As we buy and sell in the world market, technology provides salable "commodities" the production of which creates jobs here in the U.S. Technology contributes to our foreign trade balance and our ability to create jobs in three ways:

1. An industry that uses technology to improve productivity has a better chance of offering its products abroad at competitive prices.
2. Products that embody the latest technology are frequently things other nations can't buy elsewhere.
3. We add to our income by selling our know-how in the form of patents, franchises, and manufacturing rights abroad.

America's technology is wanted around the world. A good example has been our aircraft. About three-fourths of the commercial aircraft in use around the world today were made in the U.S. Our technology prevails not so much in aircraft design as in manufacturing know-how, productivity, and servicing—a major span of the innovation process.

"The U.S. has been consistently selling four to five times more technical know-how to other nations than it has purchased from them," says Dr. Thomas A. Vanderslice, vice-president and group executive of General Electric. "This has been a major contribution to U.S. jobs and a positive offset in our balance of trade in recent years."

Although the U.S. has long demonstrated a strong lead in many technologies, there are signs that we may be losing that edge. The International Economic Report of the President, transmitted to Congress in January 1977, sounded warnings regarding several areas in which foreign nations are rivaling us. "During the next decade the U.S. faces new challenges in its efforts to remain dominant in international sales of computing equipment, a commodity in which the U.S. enjoys a positive and growing balance of trade that is now over $1 billion.

"West European and Japanese computer manufacturers have begun to shift their resources toward the development of small business computers, minicomputers, microcomputers, and peripheral equipment.

"Japanese manufacturers have recently announced minicomputer models that offer many of the important system features and performance capabilities of the newest U.S. machines."[5]

The quest for innovation comes from the need to offer better products to the world market and to improve our own productivity so we can compete on price. Much of the creativity that comes from industry is prompted by its own demands for innovation. Industrial customers have traditionally demanded innovation in their production processes for product improvement and cost reduction. In recent years, the added problems of meeting public standards for environmental protection and human safety have intensified the pressure.

The money needed in the years ahead for expansion, efficiency improvement, and compliance with regulations has been pushed far beyond the expected supply of funds. There is more demand than ever for new technology to sidestep this shortage. There are many cases in which a company has not only reduced capital requirements by going to a new technology, but lessened the requirements for raw materials and cut or eliminated its pollution problems.

By competing successfully with better ideas and better prices we can curb job erosion from foreign competition. Although we may not be able to export any product we want and may have to yield to foreign superiority on certain items, we can, on balance, come out even or ahead.

If we fall behind, we cannot improve our standard of living for long. And, contrary to what we may think, our standard of living cannot be advanced simply by keeping abreast or slightly ahead of foreign competition. Improvement in our standard of living, as Mr. Alstadt says, depends on the improvement in our own productivity in relation to our own record. The bigger the advance, the bigger the improvement made possible in the standard of living.

During the last decade, we have managed to add 15 million jobs in the U.S. The pace during the next decade will have to be faster to keep up with the growing number of people entering the workforce. In order to reach that pace, "a whole new round of innovation is necessary," says William C. Norris, chairman of Control Data Corporation. "Innovations are needed like the discovery of electrical power, the invention of the telephone, or the beginning of the chemical industry in the 1880s. There are many likely targets: solar energy, fusion energy, new products that use much less energy, new materials, water conservation methods and new agricultural technologies."

Dr. Chauncey Starr, president of the Electric Power Research Institute, points out that 80 percent of our economy today is based on products that did not exist 75 years ago.

Revolutions And Evolution

We are now in the midst of an electronics revolution that is not only creating new products but altering the way we make conventional products such as autos, watches, sewing machines, and appliances. It will bring us two-way cable television with many possibilities for altering work, education, and entertainment.

The next revolution is likely to be in biology. In addition to improved medicines and medical techniques, we may see giant steps forward in the avoidance of disease through the manipulation of genes. The biologists may show us the route to new foods and fuels.

Robert Colodny suggests some priorities for scientific pursuit to open the doors to new technologies: food production, plant biology, harnessing bacteria to perform work, population control, desalinating seawater, long-range weather forecasting, geochemistry, and the process by which people learn.

Intelligently applied technology can lay the foundation for a better life and give us the options to continue the evolution of man. "Greater knowledge implies a greater chance of survival—which is what evolution is all about," write Richard Leakey and Roger Lewin.[6]

Our many physical needs are reference points that can help us set a course for our technological efforts. But our psychological and spiritual needs, if we can find a coherent pattern in them, might suggest the proper heading for our evolution. We must, therefore, confront the difficult questions: "What is life?" and "What is the good life?"

PART II
Misconceptions and Apprehensions

4 Broken Promises

"I believe we are observing a race between the trend toward the successful social application of technology and the trend of public alienation with technology. I fear that if there is widespread disillusion with technology, scientific thought itself will not long be free and supported, and some dark ages may be upon us."

Clark Abt
The Social Audit for Management

Technology is a companion of freedom and a key to America's varied life-styles, yet it has acquired a bad reputation.

Early man learned to control fire. But by 1945, his "fire" threatened to go out of control. The scientist had penetrated nature's secrets to the point where he could undo the universe. The layman did not understand the secrets of the atom, but he lived in fear that, at any time, he might be the victim of "the bomb."

Man did not understand the new dimensions of technology, but realized he was somehow responsible for its careful management. As Teilhard de Chardin had written a quarter century before detonation of the first A-bomb: ". . . the man of today acts in the knowledge that the choice he makes will have its repercussions through countless centuries and upon countless human beings."[1]

Man can use technology to destroy as well as to build. He can stumble into undesirable side effects or unknown long-range consequences of his actions. But this is not a reflection on the degree of technological advancement; it has been the case from the days of the first club or stone knife—even as far back as when man relied on his empty hands.

It is quite likely, too, that since man has been man—since this creature acquired a conscience—he has questioned the use of his abilities. Although it seems to be in his nature to resist nature, he suffers doubt and perhaps guilt in doing so. How far should he go in changing himself? How much should he use his technology to extend himself? What are the ethical bounds that should not be crossed? What are the social customs that can be expected to yield with change?

A certain amount of challenge to technology should be expected as its own natural by-product. Science and technology are founded on questioning. "Scientific and technological progress itself was born of the right of free inquiry, of the right of free criticism, . . ." says Jean-François Revel.[2]

Increased attacks on technology are partly due to the serious challenges laid against our entire social structure in recent years. Technology and the institutions that deliver it—industrial, governmental, and educational—have fallen into disfavor together. But a society that is self-critical can be a healthy one if it is self-correcting rather than self-defeating. The obsolete must be replaced with the new. Revel asserts, "No economy can survive in a technological age unless it is constantly being challenged and renewed by the collective intelligence of its citizens."[3]

Were Our Hopes Too High?

In technology, we saw the promise of making all our dreams come true—wiping out illiteracy, eliminating disease, fighting crime, and on and on. But those dreams haven't been realized. We haven't used technology to combat our most serious problems or concentrated it on making the world a better one. If we are "better off," we doubt that *we* are any better.

Our failure to effect those changes and improvements is too often labeled as technology's failure rather than our own. If we owe anything to technology, perhaps it is a sense of guilt. ". . . we are being forced to face up to the fact that it is we the people who are to blame for the ills of society, and we don't like it," writes Hal Hellman in *Technophobia*.[4]

Technology really came of age during World War II. In the years that have followed, our hopes were higher than ever that technology was the answer to many, if not all, of our problems. With massive amounts of tax dollars pumped into the defense effort and space programs, we Americans probably more than any other people looked to technology with boundless confidence.

After decades wore on with defense and space projects getting most of the federal outlays for R&D, millions of people began to feel that their aspirations were being ignored. They had a long list of other interests. These dollars, they felt, were being stolen from applications that could benefit each of them more directly and quickly.

These projects did bring benefits—spin-offs in such things as communications, weather forecasting, health care, new materials, and management techniques. Ted A. Gibson, an economist with Rockwell International Corporation, points out that the lunar-landing program "required continual probing of new frontiers." It delved into all the scientific and technological disciplines, demanding a "forcing of technology, compressing into one decade the normal advances of several."

Even for those who do appreciate the spin-off benefits received or yet to come, the payoffs look like crumbs beneath the table at which defense and space have dined too well.

When a nation strains daily under severe social problems and deploys its brainpower and resources to other ends, discontent and disillusionment come as no surprise. Opinion surveys have repeatedly ranked crime as a priority problem. Yet, less than two-tenths of one percent of federal R&D spending in 1976 was applied to this area. Energy development and conversion, certainly a prime concern in the last five years, received only 7 percent of the 1976 funds for R&D. Energy research is on the increase, but it lags far behind defense despite the fact that events of recent years have made people more sensitive to energy needs than defense requirements.

"If We Can Put A Man On The Moon . . ."

Technology's great powers have been demonstrated. The know-how that built nuclear bombs can bring the world limitless energy, we were told as World War II closed on a note of horror. The know-how that put a man on the moon, we assured ourselves, could bring us blessings in everything from health to housing. If technology can do so much, why hasn't it? There are reasonable answers but the public hasn't heard them.

One of the answers has to do with what technology can and cannot do. Technologists deal with the possible. They work toward what might be. When they speak of a possibility, the time interval before realization of it may be far longer than the public listener thinks. It may even be longer than the technologist thinks.

Another reason for the so-called failings of technology may be due to its successes. Each time we solve a problem through technological innovation, there is apt to be someone asking for similar treatment of his problem. Each success elevates our expectations.

Some of the disappointment with technology comes from laying the wrong problems at technology's door. We have asked for solutions to problems that are not entirely technological. Often, we haven't defined the problem or traced it to its roots. We can, for example, develop equipment to prevent and detect crime and to apprehend criminals. But is that the problem? Perhaps we could provide a more effective solution to the crime problem by using our technology to create jobs—the right kind in the right place. But is that the problem? Lack of income is not the only cause of crime. We can use our know-how to restructure communities and to counsel individuals. But can technology alone isolate and eliminate the problems of people relationships and unethical behavior? Perhaps it can help by leading us to ask the right questions about life and relationships. But that isn't what we ask of it. That would not bring quick solutions, and we like quick solutions.

One better weapon after another has not eliminated the possibility of war. The telephone and television make it easy to transmit information, but they do not insure meaningful communications or understanding. Technology is tools. Tools are not solutions; they may only facilitate solutions.

A successful solution begins with defining the problem and depends on applying the right skills. We've heard again and again, "If we can put a man on the moon, why can't we solve the problems of the city, pollution, and energy?" As has often been pointed out, the moon project was well defined and involved only technological problems, not social ones. In addition, says Donald Alstadt, "the we's that put men on the moon were highly trained, competent scientists, engineers, and astronauts, and the we's that have been in charge of the poverty and urban problems are a completely different group with an entirely different form of education, attitude, methodology, training, and experience."[5] If technology is to be applied effectively to social problems we need blending of skills from both the technology and the pertinent social sectors.

Because of certain social trends, some of which have been aided by technology, society is making it more difficult to solve the problems that technology previously was able to handle. Growth in population and the trend to population concentration is putting increased pressure on technology, says General Electric's James Young. "This concentrates energy demand, water consumption, waste generation and disposal, pollution of air and water beyond their natural cleansing rates, transportation, communications, and services of all kinds, including medical and entertainment. In these ways, it puts a disproportionately high burden on technology to serve society."

Conflicting Choices

By bringing us more material security and more options, technology has made us harder to satisfy. It has not only elevated our expectations, it has diversified them. There is no Great American Dream anymore. There are millions of them. There is no longer a national unity as there was during the war that rushed technology to adulthood so it could serve.

Sociologists and others call this the "age of entitlement." This is an age in which wants have become rights. The wealth and freedoms built with our technology have earned no "thank you." We assume that what was in place when we arrived on the scene was always there. We want instant gratification. Since we are highly aware of our capabilities, we feel betrayed when society—or "they"—can't or won't respond to our demands, no matter how unreasonable the demands may be. This may be an age in which nothing is unreasonable. It follows the age in which we thought everything was possible.

The problem, says Lewis Branscomb of IBM, "is that for the first time . . . we have discovered . . . that technology does offer very significant tools and capabilities to address an enormous range of problems that human beings assumed in previous generations were outside their control altogether. The real problem is not how do we deal with some new threat, it is how do we deal with the expectations and the opportunities that arise from extraordinary possibilities. It is dealing with the problem of conflicting choices that strikes me as the real challenge."[6]

This is, also, the age of personhood, selfhood, self-awareness, self-actualization. William Glasser explains its origins in *The Identity Society*. His thesis is that Americans historically worked toward survival goals up through World War II. Since then, having attained a high level of material well-being, we have searched for individual identity. People now seek meaning in their lives through recognition as persons. This has changed attitudes toward work, institutions, and even the family.

The Transition Of Values

People are concerned because technology has not satisfied their wants and responded fully to their values. They are disturbed, too, by the changes that science and technological innovation work on their values. Throughout man's evolution, the intellectual awakening through education and the application of know-how has influenced his values. Science, in particular, challenges doctrine, tradition, and values as it uncovers truth and continually slides a new basis under what people believe and what they do. Values are constantly, although sometimes painfully, refashioned to reflect the truth as it becomes more clearly seen.

Our continually expanding knowledge and capabilities change the way we view the world. Technology, through its impact on people, threatens those cultural institutions that are not resilient or adaptable. It challenges national boundaries and cultural differences. In Quebec, for example, provincial officials have been troubled for years over the ability of English language television from the other provinces to erode the special culture of Quebec. They have worked to build a base of French programming to block the invasion from the rest of Canada and the U.S. Many Canadians, for that matter, have been defensive about the influence of U.S. newspapers, magazines, books, and television on their national culture.

Some psychologists and sociologists believe the trend to regionalism—as in Quebec—or the search for one's "roots" and cultural heritage is a reaction to the fear of being swept up in a technological oneness. Ironically, technology gave us the security and material wealth so we could afford to study and savor life.

The cultural undoing brought on by technology can open the way to the establishment of richer cultures. As truth and change erode the pillars of one culture we have nothing to fear and much to gain if we erect new ones based on our new knowledge. To resist this process would be to say we are satisfied with a culture based on ignorance.

We have undergone a substantial shift in our values—not only in recent years but over the history of our country. Viewed against the history of all mankind, this has been a nation much concerned with values yet one that has shown willingness to make significant changes at a fast pace.

With the highest level of economic well-being and unrivaled freedoms, Americans have been released from many of the limitations that had been translated into values or morals. When you're poor and you need food and shelter, the need to work is not hard to understand; it's immoral to do otherwise. Work becomes an "ethic." When you depend on a modest living from your farm and can't afford to hire help, a large family is a necessity. The value of a marriage increases with the number of children it produces. Marriage and parenthood, therefore, take on numerous moral stipulations. When the economic bases for these values are eliminated, our morals are free to change so long as they stay within what we perceive as more universal ethical bounds. New technological capabilities challenge us to distinguish alterable morals from inalterable ethics.

"Our attitude toward sex, our attitude toward birth control, and our attitude toward human behavior have probably been influenced, for good or bad, more in the last ten years by 'The Pill' than any other single influence in the recent history of man," Donald Alstadt points out.

We may or may not obtain a more ethical or higher quality life from a higher material standard of living and the options laid before us by technology. The quality of life depends on inner personal well-being, interpersonal relationships, and aesthetic and spiritual considerations.

Jacob Bronowski believed that science humanizes our values—that it, at least, has the potential for making life better in the ethical sense. "Men have asked for freedom, justice and respect precisely as the scientific spirit has spread among them. The dilemma of today is not that the human values cannot control a mechanical science. It is the other way about: the scientific spirit is more human than the machinery of governments."[7]

Alstadt sees demonstration of this in the events of recent years. "Technology has produced a generation that is more skeptical about war, more enthusiastic about the possibilities of machines, more tuned to computers, less desperate about accumulating wealth or conspicuous signs of wealth, more open to ideas, more tolerant of eccentricity, more committed to sweeping social change, less concerned about what it can get out of life, and more concerned about how it can make the system work."

Blindness To The Potential

As our values shift, the possibility arises that they may create an atmosphere which works against the advance of science or the application of technological innovation.

H. P. Rickman of the University of London warns that the dynamic spirit that sustained technology might be blunted by technology's success. As people become more affluent and feel less compelled to work hard, he says, they may put less value on ingenuity and the spirit of adventure. Technological progress could come to a halt as people become unwilling to make the necessary social adjustments to accommodate it.[8]

The predominant ethic throughout the history of technology was one which equated the good life with dedication to one's duties. But today, with the emphasis on the exercising of one's rights, we may demand more from technology than we put into it.

The drive for equality, for example, could be a depressant to our innovative capability. While we should strive for equality in the dispersal of the fruits of innovation, we must recognize that the ability to innovate comes in unequal measures. We cannot obtain creativity by "averaging-out"—in education, compensation, or recognition. In the search for truth and beneficial change, all ideas are not equal.

Our values greatly influence our receptivity to technology. Fletcher Byrom fears that we may be entering a period in which we fail to use our technology properly because "our value system is such that it is based on knowledge which is obsolete . . . and, therefore, society cannot perceive the potential that technology offers it."[9] For instance, he says, we may engage in featherbedding and other methods of creating employment when "machines could be doing the work and those people could be doing something more useful for themselves."

The values we express in the marketplace or through the political system determine, to a large degree, the uses to which technology is applied. While we are aware of the cataclysmic ways in which we might use technology, we sometimes overlook the harmful effects of using technology as a palliative. If we treat the products of technology as ends in themselves—if we think that a higher material standard of living is *per se* a higher quality of life, we have gained little from our know-how. We can sink into a purgatory of materialism. Products themselves do not constitute materialism, but our attitude toward them may. If their possession and use becomes an end in itself and we embrace them as an escape from life, that is materialism. Floating along on a hedonistic trip prevents us from savoring and finding meaning in life as surely as a hailstorm of nuclear bombs would. We have to distinguish between "wasteful materialism" and "humanitarian materialism."[10] Since we have not done a good job of making this distinction and have found all too many people receptive to materialism, both as buyers and producers, we have provided justification for some of the antitechnology feelings.

The shift in values caused by the uses of technology has brought us to a point at which we can see great potential for advancement but also the strong possibility of regression. During this time of crisis there are three major alternatives for the future of technology, says Clark C. Abt, a social scientist who has studied the responsibilities of organizations in society. They are: the possibility of increasing public disenchantment with technology and reduced support of it, polarization of opinion which would likewise result in the loss of potential applications of technology, and redirection of technology to urgent social roles.[11]

The fires of innovation may be doused with antagonism or allowed to burn out as the result of apathy or polarized opinion. Until we clear up the misconceptions about technology's impact, we will not be able to direct it effectively toward fashioning the foundation for a better life.

5 Too Much Growth

"We are all afraid—for our confidence, for the future, for the world. That is the nature of the human imagination. Yet every man, every civilization, has gone forward because of its engagement with what it has set itself to do."

Jacob Bronowski
The Ascent of Man

"Haven't we had enough technology?"

Many Americans question the need for more technology just as they question the need for further economic growth. They see both forms of growth in negative terms—generating pollution, using up resources, and creating inhumane systems. In their concern for the quality of life they regard technology and economic growth as opponents.

Perhaps they have overlooked the fact that many of our problems have been lessened by technology and that others are tradeoffs for problems that were far worse. We suffer emotional problems and disease, but we live twice as long as our forefathers of two centuries ago. Today we have smog; yesterday we had manure and horse carcasses in our city streets.

Concern over growth has been fanned into crisis proportions over the last two decades. A book called *The Population Bomb*[1] underscored our fears that we will soon blindly expand our numbers past the limits of our social institutions and physical resources.

The book, *The Limits to Growth*, which grew out of work sponsored by the Club of Rome, warned that the world was pressing its physical limits in terms of resources and its ability to absorb more pollution.[2] While the Club of Rome may not have intended to set itself up as the spearhead for no-growth thinking, its name became synonymous with that concept.

Problematique

Aurelio Peccei, Italian industrialist and founder of the club, says he became concerned with the intertangled economic, technical, political difficulties around the world which constitute a "problematique." Each problem is related to other problems and the solution to one may aggravate others.

In 1968, Peccei brought together a group of scientists, sociologists, and economists to discuss the problematique. Convinced that they had to dramatize the object of their concern in terms that could be widely understood, they found a vehicle in the work done by Jay Forrester, a professor at MIT who had developed a rough computer model of an interconnected world system. The club sponsored research to extend this work in a project headed by Dennis Meadows, a Forrester assistant. Projecting such things as resources, food supply, population, investment, and pollution, the team concluded that population and economic expansion would push us to the limits of our finite planet within a few generations.

Peccei, in his own book *The Human Quality*, says the report was intended to show that there is time to avoid disaster if growth is restrained, regulated, and redirected. The people who accused it of advocating zero growth understand nothing about the Club of Rome or of growth, Peccei says. "The notion of zero growth is so primitive . . . and so imprecise, that it is conceptual nonsense to talk of it in a living, dynamic society."[3]

The No-Growth Debate

If *The Limits to Growth* was wrongfully attacked for advocating zero growth, it was also wrongfully supported by many on the same grounds. Already enjoying swelling support in many quarters, advocates of zero growth now had their "bible." Whatever their objectives, conservationists, opponents of war, birth-control advocates, and others now had a "scientific" proof to support their cause. Many causes fell into line under the banner of no-growth.

The finiteness of the world's resources was simple enough for anyone to grasp. Without really defining what resources are or examining how technology can extend and create them, one could expect consumption to outstrip supplies. The conquest of nature—once a source of pride for man—had become a grave concern. The unfairness of a few nations steeping themselves in material goods at the expense of billions of people at or below subsistence level found sympathy in a nation already reexamining its role in the world.

Zero growth became a popular catch phrase. Some people thought it made sense. Others found it a means for winning points for their own special cause. Few really got the message about the problematique.

A number of people argued that the no-growth scare was ill-founded. The world's problems, they maintained, would be solved through economic growth and technology—not by rejection of it. They pointed out that economic growth had not been slowed by lack of resources in the past because improved techniques for mining, processing, manufacturing, and materials substitution had always pushed forward resources "limitations." The earth's supply of resources is virtually limitless, providing we are willing to apply the technology and pay the price. Resource discovery and development have responded to demands.

Likewise, these people said, environmental concerns demand more technology—not less—in order to clean the water and the air. The pollution control equipment industry was growing fast, and many industrial centers could point to cleaner air and rivers despite continued economic growth.

Zero growth seemed to be of no practical value to those working directly on solutions to social problems. How could the U.S., for example, create more than a million new jobs each year without economic growth? If our economy were to plateau or fall back, how could we share with poorer nations the products of our know-how? Within our own borders, experience should have taught us that the less fortunate suffer most in economic hard times. Over the long pull, technological and economic progress have worked toward an equalization of well-being in the U.S. despite the fact that our society makes no promises of equal sharing of its wealth.

For anyone favoring a no-growth situation, Fletcher Byrom of Koppers Company has a suggestion. "Visit a crossroad in Bangladesh, a slum in South America, a village in Africa, and announce, 'Good news, friends! We have just decided on a policy of zero economic growth that will freeze everything where it is.'" Mr. Byrom, the cheerful corporate philosopher, will notify your next of kin.

At a conference called "Limits to Growth 1975" in Texas, nearly 300 futurists, writers, industry leaders, and government officials from around the world met in October that year to discuss the problems of growth or lack of growth. From Iran, the ambassador-at-large and chief of that country's economic mission said: "The Third World is not opposed to economic growth either for the poor countries or for the industrial world. As a matter of plain common sense, the only way the poorer countries can ever catch up with the richer nations is to grow a lot faster, continue to consume a lot less, and invest a lot more." What the poorer nations do object to, he went on to explain, "is unbridled and self-serving growth on a national scale at the expense of orderly global progress and development."

Did We Hear The Message?

The debate over growth had called attention to a number of worthwhile observations: that we do face global crises, that no nation is an island unto itself, that progress has not been equally shared, that what has been called "progress" may not be what people want, that we must increasingly view our actions in light of their long-term and global implications.

The zero-growth movement had enjoyed a certain moral appeal from the start. But people began to realize that the question was not whether growth *per se* was ethical. The question centered on how we can direct our economic growth and technological ingenuity to serving the world today and tomorrow.

Opinions of most futurists have begun to settle somewhere in the middle. The Club of Rome's second report—*Mankind at the Turning Point,* published in 1974—made a distinction between good and bad growth. We cannot permit more "undifferentiated growth" or growth for growth's sake, it maintained, but we should have "organic" or selective growth. "For the first time in man's life on earth, he is asked to refrain from doing what he can do; he is being asked to restrain his economic and technological advancement, or at least to direct it differently from before . . . He is being asked to concentrate now on the organic growth of the total world system," the report read.[4]

Americans had put a lot of faith in economic growth. Steadily increasing wealth had meant economic betterment for virtually all. But some people were slowly beginning to recognize that growth can't solve all problems and should not be used as a palliative. Further growth, they began to fear, might mean unacceptable social costs. Most people were concerned that their system was in jeopardy and that they might not be able to get all the things they wanted. They did not understand the causes of the turnabout, and they certainly did not believe that they themselves were part of the problem. The man in the street had not yet gotten the message about a problematique; people, by and large, had not learned to take a global view of things. If they had, they would have been seeking ways to change their daily living patterns for the good of the world. Instead, they saw a challenge to their patterns—a conflict.

It would be hard to imagine a more dramatic warning about the limits of our current resources than the oil crisis of 1973–1974. The message should have got through as Americans waited in long lines at gas stations. True, the oil crisis was not caused by a drying up of Middle Eastern oil. The shortage was a political-economic one. Middle Eastern leaders know their supply of petroleum could run out in the next three or four decades; they want to earn a fair return from it while they have it and while the world still wants it.

The Club of Rome's message, had we listened, is that our limits to growth are not so much physical as they are cultural and economic. Many of us chose not to regard such a shortage as real. "There's no shortage of oil; the problem is created by the oil exporting countries." But a shortage is a shortage. Even worse, many Americans—some businessmen included—blamed the crisis on the oil companies who were, they asserted, only trying to boost their prices and their profits.

In the years since the oil crisis, oil consumption in the U.S. has risen past the precrisis levels, and imports provide an even larger share of the supply now.

During the winter of 1976–1977, when many U.S. cities were crippled by abnormally low temperatures for a prolonged period, supplies of natural gas reached crisis lows. But, once again, there were skeptics. Many people admit defiantly that they did not turn their thermostats down when asked to conserve fuel. This crisis, too, was contrived, they feel. In addition, one might wonder how many people would keep their homes at a high temperature even if it meant robbing someone else of heat altogether.

As yet, we have not resolved the differences between our national conscience that is beginning to reconsider the consequences of growth, and our individual wants which still push for growth. The U.S. is not alone in this schizophrenia. Sweden, for example, a nation with a higher standard of living than ours, has heard the zero-growth arguments. But nationwide surveys show that Swedes—young and old—want more cars, energy-consuming appliances, and one-family homes.

Conflicting Demands

Widespread pressure for reform on the national and even global scale coupled with widespread apathy toward reforming our personal lives has put businessmen in an untenable position. They must meet increasing demand for goods and services within newly imposed restraints. We ask, as a nation, for appliances that are energy efficient; on the salesfloor we show little interest in such features and look for the more exciting options. We force our auto industry to build cars with better and better gas mileage; as individual customers, we threaten to ruin the average mileage of the cars they sell by buying too many of the heavy users.

Voluntary Simplicity

Some people have revised their personal priorities, but we are a long way from making a significant shift. Millions of people are practicing, to varying degrees, a life of simplicity, says Arnold Mitchell, senior social economist at the research organization SRI International. They are buying less, doing more for themselves. The products they buy tend to be durable, nonpolluting, simply made, and energy-efficient. This trend to "voluntary simplicity" could conceivably, within about 50 years, affect the life-style of half of America's consumers, believe Mitchell and some of his colleagues at the giant research firm. That would be significant.

Perhaps this trend will bear out what Max Lerner asserted in the fifties: "The vast array of available commodities has become an American way of living, but it does not follow that Americans are more likely than others to confuse living standards with life values, or mistake good things for the good life."[5]

We might say the growth debate is leading to a victory for all of us. We have begun to think about where we are headed. We have begun to realize that growth as we have known it faces physical, economic, political, and moral roadblocks. We have begun to turn our attention to the definition and purpose of growth and the direction of our technologies.

It seems quite clear that we can no longer project ourselves forward as a "consumption society." We are already in the process of becoming a conservation society. We will change the way we live, the way we work, and the way we do business. "We in America are learning from harsh experience that while it may still be true that we can have anything we want, we can no longer count on having everything we want all at the same time," says Fletcher Byrom. "We must choose."

We are "entering a new age," says Dr. Sherwood L. Fawcett, president of Battelle Memorial Institute. This is the beginning of an "age of conservation"—a development with the power of a second industrial revolution. "We're in a closed system," says Dr. Fawcett. "Nothing is free. Our technology was based on an open system—free air, free water, etc. Now, the job is to retool toward a closed system. There will be all kinds of discontinuities, but it's not dismal."

Appropriate Technology

In the coming years, we will hear more and more about selecting "appropriate technology" as we strive for organic or rational economic growth.

Between the extremes of technology for the sake of technology and no technology at all we can sort out the technologies that will serve us best and employ them to achieve the ends we want.

British economist E. F. Schumacher started the world thinking about "intermediate technology" several years ago—not carrying technology to its ultimate but selecting the best from it to suit the situation. An undeveloped nation, for example, does not gain by leaping to the newest forms of automation. That would minimize the number of jobs (when labor is plentiful). Furthermore, the people are not likely to comprehend complex equipment or feel comfortable with it.[6]

Speaking at the Limits to Growth 1975 meeting, two years before his death, Schumacher denounced the growth debate. "To talk against growth or to talk in favor of growth is emptiness," he said. "So we shall speak now about real things and one of the most real things is technology." He denounced, also, those who speak of whether they are in favor of or against technology. "I will have nothing to do with any such discussion. I am looking for an appropriate technology to deal with some of the problems which are not getting easier but rather more difficult."

Support for his concept, said Schumacher to the surprise of many, "comes normally from people in big business. Apparently, they have to rub against reality all the time and they can understand these things. Strange to say, the intellectuals, who ought to be the most wide-awake, get flustered, get annoyed, and ask strange questions."

Richard Tullis, of Harris Corporation, says, "We should give up our silly living patterns and devote our technology to how we want to live." Industry has to devote its technology to something beyond labor savings, he warns. The U.S. shows signs of becoming a labor-surplus nation so labor savings make less and less sense economically or socially.

Many business leaders, particularly those in large multinational companies, are well aware of the world's social, political, and economic pressures—the problematique. They are aware, too, that we have entered a period of slower economic growth. Even if we wanted to, we could not expect to ride the winds of fast growth as we have done since World War II. Most economists expect that growth in the last quarter of this century will be at about half the rate it was in the quarter century after the war. A slower growth rate will allow less room for error. It will necessitate more careful setting of priorities so that we can, at least, attend to our most important wants.

Pressures for greater humanization of the delivery systems for products and services and in our jobs will necessitate a change of direction for many of the technologies employed in industry. Some managers are already aware that bigger isn't necessarily better. While the trend in technology has generally been to bigger systems and larger manufacturing operations, they question the effectiveness (as well as other values) of extending this pattern. They suspect that the demand for quality and better service will necessitate smaller scale operations that allow for improved person-to-person relationships and more meaningful work. The route to improved productivity in terms of quality and quantity may lie in making the machine better serve the worker. Both here and abroad, even some of the largest companies are experimenting with limiting plant size and designing more content into jobs in the hope that more meaningful work will be matched with better performance.

Voluntary simplicity, appropriate technology, slower economic growth, humanization of work—none of these call for the end to technological innovation. On the contrary, they call for more of it. They call for reversal of some technologies and the acceleration of others. They do suggest that we are in a period of transition. Technology will have to be better evaluated and better managed. There will be less tolerance, demand, and means for producing innovations that do not contribute to basic meaningful needs of man. In our efforts to make the best of our resources—including our know-how—we need a climate that stimulates creative problem solving.

6 The Perception of Change

"And I asked myself about the present; how wide it was, how deep it was, how much was mine to keep."

Kurt Vonnegut Jr.
Slaughterhouse-Five

Change is coming too fast, many Americans fear. They feel they cannot or, at least, would prefer not to cope with more of it. Some of them are troubled by fears that change is eroding our society.

Change has come faster and faster. But the impression of accelerating, unrelenting change needs examination. We must study the pace up to now, the likelihood of continued acceleration, and the nature and causes of change.

If we condense the history of man into one day, we would find that nothing much happened during the first 23 hours and 54 minutes. It was 11:54 P.M. before man graduated from using simple stone tools and wandering in search of food to begin the simplest forms of agriculture. More than three minutes passed before he learned to use metals. It was practically 11:58 before he began to write.

Until that time, man's concerns had been confined almost totally to the present. There was no record of the past and little opportunity to do much about the future.

Only in the final seconds before midnight did the culture and civilization of today begin to take shape. The clock ticked off these major technological developments just before midnight:

11:59:43 gunpowder
11:59:47 movable type
11:59:55 the steam engine
11:59:58 the automobile and the airplane
11:59:59 electronic computers
11:59:59 awareness of and concern for environmental limitations.

It's understandable that we might be frightened by what has happened in the last "minute." It is midnight now, and we wonder if there'll be a tomorrow.

If we were to plot the rush of change in the last minute of our first day and extend it through the second day, the straight-up trend would represent change coming so fast that it would exceed man's ability to comprehend it, much less control it. But technological innovation is not some outside force. It moves no faster than man moves it. It is, after all, the work of man.

This is not to say, however, that an individual cannot be flooded with more change than he wants. Response to the diversity of wants in a free and affluent society creates a cumulative situation in which each person may be surrounded by more changes than he would like. We do not share the same wants, we do not share appreciation for all of the solutions, but we have to live in the midst of them.

Limits To Acceleration

Some of the basic avenues of technological innovation may have peaked out in their rate of change—if, in fact, they offer any significant growth at all.

We can see this in the speed of travel, for example. A century ago, a fast moving horse might enable man to travel 20 miles per hour. By 1900, a few people were able to approach the 100-mile-per-hour mark in trains and, later, in autos. Fifty years later, man's top speed was in the neighborhood of 1,000 miles per hour. In the 1960s, he had attained a speed of 25,000 miles per hour. At this rate of increase, he could expect to be traveling at the speed of light in a century or less. What then?

Half the world's population has not yet traveled at 20 miles per hour. Even in the advanced nations, many have not traveled 100 miles per hour in a car or airplane. Most seasoned air travelers haven't moved at 1,000 miles per hour; relatively few have ventured on the limited number of supersonic transports. And the men who have traveled at 25,000 miles per hour are so few their names appear in reference books.

In the speed of travel, as in other things, man's capability is a limit reached only by a few—not by all. Our personal lives and our culture are shaped by the lesser speeds. In turn, our speed is determined by our perceived needs as well as by economics. Both our needs and the cost of major advances in travel speed may be topping out.

Even if the need were felt, we have to wonder how much faster man can afford to travel. Henry Ford, who put millions of people in the 20 to 100 miles per hour bracket, worked alone in a kitchen, assembling his first auto from parts bought with his modest weekly earnings. The Wright brothers performed aeronautical engineering and research in a backroom of their bicycle shop, using homemade testing equipment. The aircraft that broke the sound barrier, however, were developed by large companies backed with government funds. Space vehicles that broke earth's bonds were made possible only through multibillion dollar programs employing millions of workers in a gigantic complex of government agencies and private companies.

To see the effect of need and economics on the pace of development we might also look at man's use of energy. By the time he learned to heat his home with coal and make use of water and wind power, he used more than a dozen times the energy consumed by his primitive ancestors. Today, industrial man has multiplied his energy consumption another twelvefold. Project that trendline into the centuries ahead and then try to think of enough uses to account for that much of an increase in energy consumption. In recent decades, our economy has grown faster than our energy consumption. Although the individual uses an increasing amount of energy in his home and auto, industry's consumption has declined in relation to its output of goods and services because of a shift from emphasis on heavy manufacturing with its high energy demand to light industry and services. The curve in U.S. energy consumption suggests that a highly industrialized, or postindustrial, society begins to relax its acceleration in demand for energy per given amount of output or per person.

More Changes, Less Impact

When people are asked which technological innovations are having the most impact on society today, they are likely to name television, computers, and new medical devices. All these "recent" developments are inventions of the 1930s and 1940s! Even the cardiac pacemaker had its start in 1932, but we overlook the 15-pound "first" and focus on today's lighter, smaller version.

Our impressions of change are blurred by the fast turnover of products with short life cycles. We are bombarded with "new and improved" versions of the same old thing in many cases as producers scramble to offer new products without venturing into significant development efforts.

But displacement innovations—those that represent sharp departures from the past and lay the foundation for whole new industries and new ways of life—are not coming as fast as they once did. The base for such innovations is eroding as we devote less money and manpower to research. During the sixties and early seventies, companies concentrated their innovation efforts at the end of the line—not new concepts but refinements of existing products and adaptation of proven ideas to new applications.

Although we cannot fairly judge the ultimate impact of the more recent breakthroughs, a significant number of scientists, industry managers, and government experts agree that greater impact was triggered in the late 1800s than we are likely to see coming from the late 1900s. Imagine the revolutionary impact of the first telephones in the closing years of the last century. For the first time ever, men could talk directly to one another over great distances. Radio soon brought regions, nations, then the whole world together as an audience that could share information and experiences. The auto gave man mobility far beyond anything he had experienced before. First, it served the wealthy as a toy; then it became a practical tool for the majority, freeing them from dependence on rivers, canals, and railroad lines. Such changes affected people's daily lives more dramatically than a moon landing or supersonic travel or even electronic Ping-Pong.

Several scientists and science writers have attempted to define some of the most significant inventions and determine their date of origin. Strangely enough, even in the lists which show how the pace of change has intensified over the centuries, a down cycle appears in the most recent years. Generally, the number of inventions in the last quarter of the nineteenth century and in the second quarter of this century is greater than for the quarter century just ended. The period from 1875 to 1899 that gave us such high-impact inventions as the telephone, filament lamp, automobile, and radio was a prolific time. From 1925 to 1949, another surge of invention brought us television, the nuclear bomb, and the computer. The period from 1950 to 1974 looks considerably less innovative.[1]

The Eye Of The Beholder

"The perception of a speed of change that is outpacing the capability of society to keep up with it is hardly a new one," says Hellman.[2] He also suggests that "future shock may be a symptom of our egocentricity—the feeling that everything of importance that ever happened has happened right here and now, or at least within our lifetimes."[3]

The future shock that Alvin Toffler described in the book that popularized that term was caused by social upheaval and transiency. "When diversity converges with transience and novelty, we rocket the society toward an historical crisis of adaptation," he wrote.[4] The ephemeral, unfamiliar, complex environment threatens us with "adaptive breakdown." This is future shock.

Future Shock was published in 1970 on the heels of the surge of both the commercialization of many technological innovations and a host of cultural changes in the fifties and sixties. "Future shock is something that happened a long time ago," says Marshall McLuhan. Likening the foreseeable future to the scene in a rearview mirror, McLuhan says the future of the future is now—"not the cars that have gone past you but the big cars that are coming up. Future shock is, in that sense, a very distorted image of what's going on."

Author and lecturer Max Lerner says, "The danger of technology lies not so much in what its social effects are going to be—the future shock and so on. I think that was probably overstated. I think its danger lies in its impact on our value system."

We may be suffering from value shock or attitude shock—not technology shock. Although it may seem that technology has advanced faster than our values and our social structures, the reverse could be true.

John Platt of the Mental Health Research Institute, University of Michigan, pointed out at the World Future Society conference in 1975 that television and education are "great revolutionaries" that have given us fast changes in attitudes. "The speed of attitude change is now so high that the old sociologist's idea of a 25-year 'social lag' in catching up with technological change is obsolete. Today we have something more like social choice and 'social advance,' with a 'technical lag' because of the long research-and-development times required to satisfy the new social demands."

While we are blaming technology for our ills, we are more dependent on it than ever to meet our diverse demands. The pace of technological innovation needs to be stepped up to maintain cadence with social advance.

Far too many have reacted in just the opposite manner from what Toffler was urging. Rather than understanding our malaise and working on its cure, they utter the term "future shock" and slump back, convinced the future is destined to be overwhelming. Rather than molding the future, they have copped out.

If we assume the future is merely to be endured, we try to hold back its rush upon us. From that follows our resistance to scientific exploration, technological innovation, and institutional change. We imprison ourselves in the present. Most people are locked into niches from which they never hope to escape, Toffler says. Despite all the antitechnology rhetoric, "it is precisely the super-industrial society, the most advanced technological society ever, that extends the range of freedom."[5] He points out that those who would break from their imprisonment "would do well to hasten the controlled—selective—arrival of tomorrow's technologies."

Positioning For Change

"Too often, we look at the future as being a series of events, opportunities, and disasters that overtake us and engulf us from 'out there,' " says William J. Crockett, vice-president for human resources of Saga Corporation. "But we should think of the future as being ours to create, ours to fashion, and ours to dream. Our futures will come out of the best that our logic and our rationality can create. But even more importantly, they will also come from the best that we can dream."

People who find the future a nightmare rather than a pleasant dream do so because they take a passive approach to the future. Yet, in the face of complexity and change, who can feel that he is really in control?

People feel more comfortable with change if they can see the need for it and especially if they can share in shaping it. "Unfortunately, the innovator contributes unnecessarily to the changes imposed on people by his tendency to stress the novel aspects of his work," says P. R. Whitfield, a member of the Council of Engineering Institution in Britain. The introducer of an innovation would do better to "reduce the mystery and newness" and present it so that "it is seen as just a step in the natural progression of accepted patterns of work and living."[6]

At the risk of getting ahead of the social scientists, one might speculate that a strategy similar to this is embodied in the decor of thousands of retail stores and restaurants today. Their electronic cash registers and stocks of new and improved products are placed in settings modeled after the eighteenth and nineteenth century—not the twenty-first. Plastic "stained glass" and synthetic "timbers" strive to bring back something we might have preserved.

Another speaker at the 1975 World Future Society meeting, Daniel Gray, senior consultant with Arthur D. Little Inc., pointed out that change can be frightening or exhilarating. "I used to be afraid of airplanes and comfortable on roller skates. Now, I fly to a meeting such as this on a jet, but I'm scared to death of roller skates. It's not the rate of change (speed) that bothers me but my positioning."

Perhaps it is not so much technological change but social change that we are afraid of. After all, Dr. Merrifield at Continental Group reminds us, "Our society is acclimated to technology like no other."

Xerox's Jacob Goldman suggests that when people talk about being antitechnology, they should be challenged. "They'll look at their electronic watch, go home and watch TV, and use a little calculator to see how the supermarket costs compared—but they 'hate' technology."

Quite likely, we fear the social changes and revision of values that technology ushers in more than its hardware. New knowledge itself is threatening to anyone unwilling to learn and to adapt.

"We can expect trauma; it's hard for us to change," admits Dr. Merrifield. "But the opportunities are so great relative to the dangers." He looks ahead with the conviction that we can be "neither optimistic nor pessimistic; we have to be positive."

7 Planet in Peril

"It is the business of the future to be dangerous; and it is among the merits of science that it equips the future for its duties."

Alfred North Whitehead
Science and the Modern World

Mind-boggling technological disasters wait in the wings at all times today—nuclear bombs, radiation poisoning, destruction of the ozone layer.

Obviously, technology has brought us hazards. It brings undesirable products, harmful side effects, and unknown long-range consequences. It can be used to destroy as well as to build. Harm may be deliberate or accidental.

The atomic age was launched with horrifying weapons. Even their peacetime offspring frighten millions of people who are not convinced that a nuclear power plant isn't just a bomb that might go off at any time. Those who understand these facilities a little better point to the ever-present possibility of radiation leakage. And the best of the technologists debate about what to do with nuclear waste materials.

There are techniques and materials available from which to create safe storage systems for nuclear waste. But our present solutions would render the material inaccessible if it should ever be needed for reprocessing to make it into fuel again. On the other hand, we cannot decide whether we will ever want to reprocess it.

The next "atomic bomb" could be the DNA molecule. Genetic manipulation promises to open doors to the elimination of disease and possibly to improving man's body. But, at the same time, mishandling this emerging knowledge could set loose horrors by the hundreds. We must proceed with careful investigation, yet we fear accidents might occur even in exploratory research.

Not all technological danger comes from big installations or laboratories. We found that each of us may contribute to the danger by using aerosol sprays for paints, hair care products, or deodorants. Who would have thought that the propellant for the spray aimed at one's armpit might end up in the ozone layer (which many of us hadn't even heard of) and cause changes there that could lead to dangerous radiation reaching us from outer space?

We have learned, too, that some of the simple things we do can bring great personal harm. Remember when the primary danger from cigarette smoking was that it would "stunt your growth?"

The conqueror of nature is humbly realizing that he is still part of nature. He can describe nature's truths in laws and follow them for his own ends, but he cannot break them. Those which he does not understand thwart him in unexpected ways. Since ignorance of the laws can be disastrous, he is forced to use his scientific and technological capabilities to learn still more.

Truth And Consequences

We can live better within the laws of nature as we come to understand them. During the last two decades, we have become more conscious of the interconnectedness of nature. "Ecology" is a relatively new word to most of us. We realize that each entity, each action, is part of a system and that systems relate to one another in a complex whole.

We are beginning to understand the complicated relationships and processes that permit the earth to support life; we realize that this planet once did not.[1] What we once thought were independent actions have sometimes shown obvious impact on other parts of the system. Sometimes, however, their consequences are not so obvious—they require careful study over a long period of time. We frequently have to look at a problem and painfully trace its origin—from consequence back to cause.

We realize now that a technological development has to be considered in terms far beyond the direct benefit we see in it. In the U.S. we have been masters at spotting the immediate, direct consequences of a technological development by finding a market for it. We have, also, done well at accelerating our progress by finding additional uses for a new capability.

Attention to the next level of consequences came more slowly. After we reached a certain level of affluence and security, we became sensitive to the harm that we had wrought on the environment. For most of our history, we assumed that industrial might and pollution had to grow together. We now jokingly refer to the days when smokestacks signaled good business as they blackened the sky. In just two decades, we have moved from an emotional outburst against pollution to taking steps that have measurably cleaned our air and waters. We had much of the technology on hand to control the emission of pollutants, but our economic system had put no price on protecting the environment. As producers and customers, we were out for the lowest cost. Air and water were free, so we thought. The price we paid for our products did not provide for making them in a manner which wouldn't pollute. The nominal costs we paid for sewage treatment didn't do the job of protecting our streams and lakes. Now, we are applying the force of law to build the cost of a clean environment into our prices. We are thereby making it attractive to develop still better processes for controlling our effluents.

We have learned that man need not always create waste as he creates something of value. Long ago, we had demonstrated this with advanced agricultural methods. Modern farming methods—as productive as they are—have been far less damaging to the land than primitive methods. Technology is enabling us to make what we want, make it better, make it with better resource utilization and less waste, and make it in facilities that are safer and more humane as workplaces and neighbors.

We can correct for the long-term consequences of technological innovation only so far as we are aware of them. When scientific investigation raised the suspicion that the fluorocarbons used as refrigerants and aerosol propellants jeopardized the ozone in the stratosphere, we began to switch to substitutes. Ironically, fluorocarbons were developed in the 1920s as a replacement for the dangerous refrigerants then in use! They were directly responsible for saving an untold number of lives and yet they themselves may present a hazard.

We have rejected some technological innovations when we didn't like their potential consequences. The U.S. backed off from a program to build a supersonic transport, for example. Although we had the capability to build an SST, we said it wasn't worth it in terms of economic or environmental costs.

Actually, we had not made a careful study of the trade-offs. Some people were concerned with the noise levels of the plane. Others feared the effects on the upper atmosphere. Some objected to a large expenditure of federal funds to develop a product that would serve relatively few people directly. Years later, thousands of people demonstrated in protest against the landing of the European SST. Their primary complaint was the noise levels produced by the plane on landing and takeoff. Some tests showed that the plane was noisier than other jet aircraft; others, that it was quieter. One crucial evaluation was not made: if the noise was disturbing or dangerous to people near the airports, how much positive benefit to how many people would be necessary to offset that?

Unforeseen Social Impact

Perhaps the most difficult order of consequence to anticipate or observe is the social impact of a technological innovation. We not only have trouble foreseeing the ultimate development of the technology and hardware involved, we have to contend with the unknown degree of public acceptance and untold ways in which use of the technology might change behavior.

As yet, we are not very good at anticipating all the potential changes that a technological development might bring to our lives and our values. At the turn of the century, we could not have dreamed of the impact of the automobile on our mobility, our jobs, our cities, our environment, or our streets. We would have been even less likely to foresee the impact on our values and morals. Now, as we try to determine the long-range future of the auto, we cannot be sure how our values will take their turn at shaping the auto's future.

The environment surrounding the auto is changing, says Marshall McLuhan. He points to one value consideration that could affect the auto. "You cannot have a community on wheels because it is antithetic to the nature of neighborhood and community. When people begin to desire community and neighborhood, they will take a very dim view of the car—which is already happening. As people desire roots, they'll say, 'The car destroys roots,' and they'll get nostalgic: 'The bicycle is back. When will the horse be back?' "

A more subtle example of the social surprises posed by technological innovation comes from the medical field. Our growing population, particularly the growth in the number of golden agers, is a fruit of our medical technology. But it presents problems in housing, employment, care for the aged, and education.

Better Detection Of Hazards

While man has neutralized some of the natural dangers confronting him and increased his comforts, he has increased the danger of harming himself. This has been so throughout the history of technological innovation.

Why, then, does life suddenly seem dramatically more hazardous? Why do we now feel that we have been betrayed by our technology? Why have products and practices that we have been accustomed to suddenly turned on us?

We hear of poisons on the dinner table and chemical time bombs in our breakfast cereal, says John W. Hanley of Monsanto Company. In 1959 we had the cranberry scare and were afraid of eating tainted Thanksgiving dinners. Then came the ban on cyclamates, Red Dye No. 2, and saccharin, he recalls.

Danger is not new to us. But "we are doing more to detect the harmful effects of technology," says Max Lerner. "It's only recently that we have begun to measure the negatives," adds Murray Weidenbaum, director of the Center for the Study of American Business at Washington University.

Our ability to measure and interpret the effect of substances on our health "has outstripped the laws that regulate their use," Mr. Hanley explains. In 1958, the Delaney Amendment to the Food, Drug and Cosmetic Act prohibited any presence of a cancer-inducing chemical in foods. Twenty years ago, we could detect the presence of most chemicals only down to parts per million, Monsanto scientists point out. "At that time, our scientific understanding of trace quantities and their effect upon human health was, by today's standards, unsophisticated," says Mr. Hanley. With developments such as the electron microscope, mass spectroscopy, and neutron activation analysis, we can now spot quantities as minute as one part in a trillion, he says.

Our technology has not made eating more hazardous so much as it has enabled us to find traces of harmful ingredients that are too minute to be of harm. Yet our laws say we must ban the ingredient. That's comparable to saying that if every American had 5,000 apples and one apple was bad, we'd have to destroy all the apples.

No Future Without Risk

The past has never been without risk. There is no reason to expect the future to be risk-free.

"The goal of zero risk is obviously appealing but there is no way that technology can deliver on this contract," says Irving S. Shapiro, chairman of E. I. du Pont de Nemours & Company Inc. "There are ways to improve product testing procedures and the protocols for clearing new materials coming into public use. There is no way to certify that all the hidden hazards have in fact been uncovered and removed. Science cannot make life risk free, though that is precisely what is being asked in some of the demands heard today," says Mr. Shapiro.

Directing our technology is not a black-and-white matter of determining whether something is good or bad. We generally have to compare one set of risks and benefits with another—perhaps with several others.

"The types of risks we take are open to debate, but whether we will take risks is not open to debate," says Alan Greenspan, president, Townsend-Greenspan Company Inc. The former chairman of President Ford's Council of Economic Advisers believes we can no longer follow a course of risk-avoidance. "We've had too much in the way of quick-fix solutions without asking what the long-term costs are."

Saccharin had been under suspicion for more than fifteen years when a Canadian study showed that when fed in huge doses to rats it caused malignant bladder tumors. Acting under the Delaney Amendment and the general safety requirement of the Food, Drug and Cosmetic Act, the Food and Drug Administration announced in March 1977 that it was going to ban saccharin.

The public leaped to the telephone and the mailbox. Although a few people feared that they just might be prone to cancer as the result of having used saccharin, millions resented the loss of the artificial sweetener. They made jokes about the high dosages fed the test rats—the equivalent of eight hundred 12-ounce cans of diet soda per day over a lifetime for humans.

This was a case in which everybody seems to have been wrong. The FDA had to follow an inflexible law—an absolute one which legislators should have updated. It says, in effect, that since no one knows how much saccharin will induce cancer in a human any presence at all must be presumed hazardous. And the public, scoffing at 800 cans of diet pop per day, figured that since no one is likely to consume that much there is no chance that a human would get cancer from saccharin. Yet massive dosages in test animals is a standard way of speeding up the identification of cancer-causing chemicals.

Out of the public outcry came a lesson for all of us. Even if saccharin did pose a risk, the public could see that perhaps there was an even greater risk in not being permitted to use it. The artificial sweetener had probably saved thousands of lives of people who tended to be overweight or who suffered from diabetes.

Fleeing To Greater Risks

Safeguarding the health of people and their planet fires the emotions. As a result, we might flee to worse unseen risks as we try to avoid the apparent ones.

Mr. Hanley warns that the proven harmful effects of some chemicals could trigger fear of all chemicals. Thomas Miller of Union Carbide, adds: "If we create the impression that anything synthetic is bad, we're in trouble. And we're close to doing that." Chemicals tend to scare us simply with their names. Think about your morning breakfast. Among the naturally occurring chemicals you consumed: methanol, acetone, isoprene, and more, in your coffee; acetic acid, globulins, and butyric acid in the scrambled egges. And that harmless piece of toast? Ethyl acetate, methyl ethyl ketone, and amylose.

People who are frightened or confused aren't listening to what scientists or politicians are telling them. They didn't "buy" the warnings about a flu epidemic during the winter of 1976–1977.

It's possible that the public is a step ahead of the politicians in realizing that we cannot have absolute safety. They are willing to settle for acceptable risks. All they ask is the facts. They'll determine whether a given risk is acceptable in light of the benefits they expect.

Two sets of inputs are needed for risk decisions whether they involve an individual, society, or the environment. We need facts from experts on the risks, the benefits, and the unknowns. And we need people's values. By bringing the two together we can determine how valuable the benefits are and how much risk is acceptable.

The problem today is that we let emotions into the mix too early. The doomsayers rush in, seizing on bits of information and rallying public sentiment, before we look at all the available facts. Inputs from the too-few, experienced experts are diluted with misinformation spread by well-intentioned activists and self-serving power groups.

One thing we need to do, says Mr. Hanley, "is to replace the political expedient of outright bans with scientific programs which accurately define risk. We have taken the easy way out and said, 'there may or may not be a problem, so we'll ban it.' " Banning, as a quick-fix, precludes measuring the real world effects of an innovation. It may also discourage research into an innovation's risky aspects, so neither the originator nor his competitors are likely to develop an improved version.

Obviously, we need scientific knowledge to determine risks. We need to know more about the direct negative effects of an option, and we need to trace the possible indirect consequences in a world of interdependencies. We need cheaper and faster testing methods.

The Cost Factor

The adverse effects of technology on the environment have generally been due to attempts to use the cheapest process possible in manufacturing, utilities, and government facilities alike. Few people, if any, want to deliberately pollute our environment. The question is: How much are we willing to pay for cleaner environment?

One reason the U.S. did a poor job of environmental protection until recent years is that we weren't willing to pay for it. The public is now showing some awareness of costs and willingness to pay them—up to a point. "We're willing to make some tradeoffs to protect the environment," believes Dr. David S. Potter, General Motors' vice-president for environmental activities. Car buyers are probably willing to make trade-offs that cost them on the order of $10, but there is no evidence yet that they are willing to make a trade-off in the $1,000 class, he says.

We can have air and water as clean as we are willing to pay for. We might not have to shut down industries, but we would have to pay such high prices for some goods and services that our income would not go as far as it does today.

The closer we get to zero risk or zero pollution, the faster the costs climb. Reducing a discharge into our streams or air by 98 percent might cost a great deal yet be worth the price. On the other hand, the last 1 or 2 percent may cost many times more than the first big step without a commensurate benefit to society.

Not all environmental impacts deserve equal attention. Before we rush headlong into curing all our ills—even if we don't insist on zero pollution—we should know where to direct our efforts first. The Rockefeller Foundation financed a task force in the mid-seventies comprising business, labor, and agricultural leaders along with professional ecologists. The group explored the possibility of setting priorities so that projects could be financed economically and long-range technologies could be developed. It designated three categories of environmental impact: those that create irreversible damage and must be stopped at all costs; those that cause damage which is reversible if not exposed too long; those whose damage can be reversed even after long exposure.

The next step in pursuing this is to establish a task force of trained ecologists who would scientifically catalog the known environmental impacts and classify them into the three categories, says Fred Smith, an associate of Laurance S. Rockefeller on conservation. Mr. Smith, who has served on numerous commissions and acted as consultant to Presidential groups on environmental matters, believes that "business, labor and the government could then get together on a timetable for a long-range schedule that meets environmental needs and is economically feasible." This could move us to "a new world of environmental progress and energy production that would encourage, not short-circuit, sound economic and social development.

The more we advance our scientific and technological capability, the more we will be able to detect hazards and present them for proper risk evaluation. We can strengthen our ability to determine which technological innovations would be harmful to us or our environment. We can, also, be better able to apply technology to reducing hazards without harming our economic condition.

Without science and technology, we would be lost in the complexities and interrelationships of our world. With them, we can work on minimizing our risks. But we must be realistic in our expectations. We will go on living with risk. Du Pont's Irving Shapiro cautions: "We need to understand that the business of science is not to deliver certainties but to reduce the level of uncertainty about our options."

8 Inhumane Systems

"In the increasingly mechanized, automated, cybernated environment of the modern world—a cold, bodiless world of wheels, smooth plastic surfaces, tubes, pushbuttons, transistors, computers, jet propulsion, rockets to the moon, atomic energy—man's need for affirmation of his biology has become that much more intense."

Eldridge Cleaver
Soul on Ice

When primitive man's fire went out, he and his family slipped back to their primal relationship with nature. The more fire served him with heat, light, and protection, the more he became dependent on it. He had to learn to support, transport, and relight it.

Switch off the electric power for Manhattan, and millions of people are unable to move across town or even to escape from their tall buildings. They are imprisoned without water, ventilation, and communications. As a virgin farmland, the island of Manhattan might support a few thousand farmers. Inhabited by millions of residents, commuter workers, and visitors, it depends on sophisticated, complex, interrelated support systems. Manhattan dramatically represents our dependence on technology. It also illustrates how social and technological trends have brought men together in heavy concentration to the point at which they lose, rather than gain, identity as they give way to impersonal and inhumane "systems."

Technological innovations continually build an artificial equilibrium between man and nature, P. R. Whitfield points out. He reminds us of something we know all too well—"any sudden breakdown could bring about dangerously swift changes. Clearly, we must ensure safety and continuity of use for processes and products if people have become vitally dependent on them."[1]

We have built layer upon layer of technological improvement in the natural world that was hostile to us. Now, we must guard against failure of any of our technologies that might plunge us back into the natural state. As we select each technological innovation we have to be assured that, if we might become dependent on it, we can be reasonably sure of being able to sustain it. We must also consider its impact on both the natural and our man-made environment.

This leads us to a paradox. Man has trapped himself with his own achievements. The more he uses his special powers, the more he must rely on them. If he does not use them properly, he becomes his own captive, says Aurelio Peccei, founder of the Club of Rome.[2]

We have been aware of this paradox. Most of our concern regarding technology—at least, up to now—has been with maintaining the physical equilibrium between man and nature. In recent years, especially, we have begun to make corrections so that the two environments can exist in harmony. Unless we want to revert all the way back to the natural state, we know there is little choice but to carry out our technological responsibilities even if we grow a little weary at times. As Fletcher Byrom quips: "Technology is sort of like having a spouse who helps you with the problems that you wouldn't have had if you hadn't gotten married in the first place."

Independence And Interdependence

We can live with this paradox so long as we do maintain an equally important equilibrium—the balance in man's psychological world between the individual and society. Dependence on a man-made world should trouble us no more than dependence on a natural world. But we resent a man-made world if it is not man-related.

One of the strongest sources of antagonism against technology is that it is inhumane—that the social structures supported by technology are mechanical and impersonal.

Technology has made it easier to construct large, centralized organizations which heighten our sense of vulnerability and helplessness. We find ourselves dealing more and more with faceless organizations, less and less with people. To a large degree, the computer has aided this depersonalization. But is it the computer that needs reforming? Or is it the people and their organizations?

"What we have tended to do has been to internalize the machine and the computer," says Max Lerner. "I think we have to live *with* technology. We don't have to live by it. We don't have to make it the paradigm for our society or the metaphor for our lives."

Perhaps it's only fair that we act like machines since we have asked machines and organizational structures to do the humane things that we as individuals have failed to do. Demands and needs have not been met at the personal level, yet our society has tried to respond to them.

As we respond to human need, we tend to increase the level of complexity whether it be in our schools, hospitals, or our automobiles, says Lyle Schaller in *Understanding Tomorrow*. More complexity does not necessarily lead to more humaneness, however, he points out.[3]

As we emphasize human needs and deliver more complex answers to cope with them, we raise the level of indirect and perhaps involuntary involvement of the individual in society. "It seems to be a technological condition of modern life . . . that we engage in more numerous activities which generate costs we do not bear directly," writes Nathan Rosenberg in *Perspectives on Technology*. "The converse, however, is equally true. Members of urban communities not only generate costs for which they are not directly responsible but also generate benefits to other members of the community for which they receive no compensation."[4]

We find ourselves pulled into solutions of other people's problems, paying more than what we think is our fair share. Recipients or seekers of benefits feel they owe no thanks to the "third party" who picks up the tab. From both ends, the system looks impersonal.

It is not the machine but machinelike people and organizations who alienate us and turn us against technology. Technology can permit us to come closer to one another and thereby heighten our interdependence. But if we do not come together with caring and cooperation, technology only intensifies the problems we bring.

Large Scale Or Personal

We have tended to select technologies that increase our concentration into high density population centers. In our effort to generate wealth and distribute goods and services at minimum cost, we found that bigger was better—that there were economies in large scale operations. The best way to produce low-cost electrical power, for example, was to rely on large, centralized, power generation systems. The more we concentrated our homes and places of work within the reach of those facilities, the better our access to electrical power.

People have always tended to concentrate for a number of reasons—protection, companionship, commerce. In one respect, the technologies we have employed stimulated concentration; in another, they merely assisted social trends that called for concentration.

If we should choose to disperse ourselves more evenly throughout the country, we could apply our technology to that end. We could, for example, call upon the technologies that would enable us to generate electric power effectively and more economically for small communities or for individual homes.

In the past, we found that operations conducted on a large scale could reduce unit costs. They were economical. But the definition of "economical" may be changing. As we build in costs for pollution and other social negatives, "large scale" may no longer be synonomous with "economical." The definition depends on what kind of society we want.

The inevitable outcome of our social progress might be ever and ever greater interdependence. Longshoreman-philosopher Eric Hoffer reminds us that, to some degree, America owes its progress to people who "wanted to be left alone."[5] For most people, life has always been a matter of balancing independence with interdependence. The technologies we have selected reflect this. The riverboat, train, and subway gave us good mobility. But we developed the auto for more independence and privacy.

The electronics field promises to bring us changes that will rival the auto in terms of impact on our lives. We may be hesitant about further advances in electronics since we are so conscious of the centralized, complex systems already in place. But this need not be the course for the future. In fact, the developments presently blossoming the fastest are the small-scale, personal products.

"Items that have recently become broadly available, such as the handheld computer, electronic watch, and citizens band radio, enhance the public's feeling of participating in the benefits of electronics while not bringing with them discernible side effects," says Dr. W. D. McElroy, chancellor of the University of California, San Diego.[6]

Even the large scale electronic systems may become more personalized. It is not "blue sky" to talk about two-way television systems which would enable a person to sit in his home to conduct his work, select his groceries, and take part in college seminars with the world's best teachers. We already have the technological capability to deliver "newspapers" electronically, providing continually updated news as well as a library of reference material on tap with the push of a button.

The computer and other advanced technological systems may outlive their bad reputations. They are still in their adolescence, and people have to mature in their use of them. Few of us will build or program computers, but this need not stop us from better use of them. The inability to build or repair sophisticated products has not prevented acceptance of the auto, the telephone, or television. Although we are less and less able to understand the inner workings of the technology around us, we rely on specialists. "And there are enough of them around," says Continental Group's Bruce Merrifield. "The U.S. has a fantastic base of technically trained people," he says. Many of the people who are corporate vice-presidents in charge of data processing or management information systems were in college during the fifties or earlier. Few had worked with, studied, or even seen computers when they began their careers, but they were adaptable enough to take command of this new technology.

In industry one can see a shift in the way computers are regarded. Some managers are scrapping the dream of a huge central computer running the entire show for a company because they have learned several important lessons from their experience with data processing systems. They know, first, that not everything can be quantified and computerized. They are no longer awed, either, by computer specialists with "bytes and bits" vocabulary. They find little justification for mountains of data which they don't need, and they aren't content to receive it only when the system is set to deliver a periodic report. In the years ahead, industry managers will depend more and more on minicomputers that will permit them to decentralize their decision making. The larger, centralized computers with their banks of knowledge will respond to inquiries, providing users what they want when they want it.

Hiding Behind The Computer

Innovations like the telephone and auto brought technology to everyone for personal use and control. In recent years, some of the big technological innovations seemed to leave most people out. They could be understood and used by a selected few—the mathematician, scientist, business manager, or computer systems analyst. People could read about the developments and the ways in which they were benefiting indirectly, but the technologies were foreign to them.

The computer is probably the best example of a technology which defied popular control. Resentment boiled when people did have to come into contact with systems employing it. They were frustrated by systems designed to tell them what they owed the department store but unable to tell them why. They could not understand why they should have to spend months trying to get a system error corrected when the system was supposedly designed for speed and efficiency. The computer was helping someone, but not them.

People resented the computer when they should have resented the people who used it as an excuse for impersonal treatment. The lazy or inconsiderate salesclerk or credit department employee could too easily hide behind the computer. Those who designed and used the data processing systems were giving the computer a bad reputation. In time, they explained that the computer was really stupid and any errors were human errors. But to the man in the street, "the computer" was not a tool; it had become a master.

Making The System Responsive

Technology responds to human direction. Whether it appears humane depends on the people who direct it. If our private or public institutions use it in inhumane ways—even with good intentions—people will rebel. Unfortunately, technology may be the scapegoat.

Concern for the technology /man interface will have more influence on the use of technology in the future as our institutions respond to the public's demands. Both our technologies and our organizations will have to adapt to the insistence on humane treatment.

The youth have already protested against the giant universities which stripped them of identity rather than helping them find it. They resented a system which processed them by the number. Similar demands will be expressed by the golden agers, particularly as they grow in number and more "system" is established to serve them. The paperwork faced by a retiree or widow involved with Social Security, insurance premiums and benefits, and taxes is becoming inhumane.

Urban rebuilding has paid little attention to man's need for human contact and identity with his surroundings. Airport terminals and office buildings fit the parameters of "good" engineering and finance but sometimes relate poorly to the people who use them.

Technology is neutral. We can adapt it to our needs. We, in turn, can adapt to the changes it brings—if it meets our needs. We chose dependence on a technological environment over a natural environment, but we assumed that we would have control over technology and be able to make it do things our way.

Some technologies permit personal participation and control; others can be personalized only by the people and institutions that employ them. People may or may not tolerate technologies whose inner workings they don't fully understand. But they will not tolerate technologies—or, more correctly, institutions—that are impersonal, inhumane, or unresponsive.

9 Destroyer of Jobs

"Man does work for profit in order to avoid pain; but in a positive sense, he works to enjoy the excitement and meaning that achievement provides for his own psychological growth and thereby his happiness."

Frederick Herzberg
Work and the Nature of Man

One of the most powerful causes for opposition to technological innovation is the fear that it eliminates jobs. Technology has continually been used for this purpose in order to reduce costs or eliminate jobs that were hazardous or undesirable. The person who loses a job to innovation or who is threatened by such a loss, understandably, does not look too favorably on technology. As we add up the cases of workers displaced by innovation, however, we wrongly conclude that technology poses a threat to society by increasing unemployment.

Newspaper headlines sensitize us with reports of specific job losses. We hear about the loss of paychecks as, for example, United States Steel Corporation employment dropped 39,000—nearly 20 percent—from 1966 to 1976. Bethlehem Steel Company lost 28,000 jobs during those years. Not many of us are aware, on the other hand, that IBM added 94,000 jobs during that same time. We are not aware that Xerox's employment grew to five times its 1966 level with the addition of 77,000 jobs—or that 3M Company added 26,000 jobs and Texas Instruments, 27,000.[1] We do not appreciate the fact that the companies which added jobs are those most involved with technological innovation.

While innovation may eliminate the job for an individual looking at the short term, the lack of innovation is a greater threat to the individual, the company, the industry, and the nation. A look at the national numbers gives no support to the fear that technological innovation reduces employment. In the early postwar years, four out of ten Americans were employed. That same ratio applies today despite the growth in population. During the last 30 years, we have added more than 30 million people to the ranks of the employed. An increasing percentage of women have entered the work force.

Contrary to what we might expect based on our knowledge of specific layoffs, employment in the manufacturing sector has not shrunk. It has not gained either because this is not where the thrust of our economy is any longer. Manufacturing jobs held level at about 19 million through the decade ended in 1977. As jobs increased elsewhere, manufacturing's share of total employment slipped to 20 percent in 1977 from 24 percent in 1967.

The number of jobs in the service industries grew by 5 million to a total of 15 million. Nearly 5 million jobs were added in wholesale and retail businesses, bringing the total to 18 million there. In 1977, federal, state, and local governments employed 15 million persons—4 million more than they did 10 years earlier.

The figures tell the story of how the use of technology has permitted increased production of goods without increasing the number of jobs required to do so. This is productivity. But productivity alone could have reduced the number of jobs in manufacturing. "Productivity alone can be counter-productive to full employment," says William Norris of Control Data Corporation. "It must be accompanied by a creative force for new products and services that will absorb both the newly unemployed and the expanding work force."

Where Jobs Grow Fastest

There's more to the story of technology then. Jobs have continually shifted to those companies and industries that develop new products or employ more productive processes. In special studies for General Electric Company, Data Resources Inc. found that high technology industries create jobs eight times as fast as low-technology industries. It categorized industries according to the amount of research and development effort they expend in relation to their total output. Data Resources discovered that output per employee increased an average of 4 percent a year in advanced industries compared to only half that rate in traditional industries. In terms of job creation, this improvement in productivity provided a powerful advantage. "The rapidly modernizing industries surpassed their conservative counterparts by a substantial margin—2.6 percent growth per year versus 0.3 percent," says Data Resources.

A Commerce Department study shows that from 1957 to 1973, technology-intensive industries' output grew 45 percent faster than that of other industries. Employment increased 88 percent faster.[2]

Because of improved productivity in the manufacturing sector, we have been able to disperse a growing percentage of our workers into service-oriented industries. We have been able to devote increased attention to things that can give us a better life in nonmaterial terms—education, travel, health, and recreation. Whether these services really give us a better life or serve only as diversions and the source for new social problems is another question.

So long as we produce the goods we need and do not needlessly lose jobs to foreign competitors by being less efficient and innovative, there is little reason to worry about preserving jobs in the goods-producing sector. We do, of course, have to be concerned with the ability of displaced workers to find new work. The individual, in turn, has to be flexible enough to accept new employment. Unfortunately, the personal response to the threat of a job loss is often quite the opposite. Rather than change work routine, leave old companions, and possibly relocate one's home, the worker fights to preserve his present job.

For newcomers to the work force, new industries or new careers pose less of a problem. Thousands of high school and college students are preparing for careers in jobs or industries their parents never dreamed of. Yet they may fulfill their parents' dreams that they would work "with their heads rather than their hands."

Is there, after all, a finite amount of work for man to do? Few of the jobs that exist today could have been listed by even the most farsighted of this country's founding fathers. Work has continually taken on new meanings, too. There has been a decided shift from the use of manual to mental skills. There is a market for talents that would have to have been suppressed a century ago if one wanted to earn a living.

The Ever-Changing Definition Of Work

The Industrial Revolution may have saved Britain from mass starvation because her population had grown beyond what could be supported by agriculture. The factory could generate goods that could be traded for food. At a time when personal liberty was dawning in Europe, people flocked to factory jobs for which they sacrificed their freedom. If they were to eat, they had to segment their lives, giving a large share of themselves to the job in exchange for pay.

"Going out to work began with the factories about 1800," Marshall McLuhan notes. "Up to that time, people did all their work at home. For hundreds of years, work was all done at home, somewhat as the farmer still does."

As industrialization progressed, jobs in the plant and the office were segmented into still finer pieces. Early in this country, Frederick Taylor introduced a concept of work based on scientific analysis of the specific functions performed in a particular job or process. Jobs were then structured to maximize the efficiency of the worker. This often meant he performed repetitive, simple tasks.

Segmentation of man from his home, from the rest of his life, and from meaningful work was dehumanizing. Unfortunately, too many of us have outdated impressions of what industrial work is today. We may recall pen and ink sketches of the industrial revolution plant or the movie images of mindless jobs on an assembly line. But less than 2 million workers in the U.S. today are on assembly lines. Working conditions are significantly improved in the typical plant. Work schedules have yielded many hours per day and days per year so the worker can construct a life off the job. And, increasingly, the job itself is being made more meaningful.

In the years just before and after World War II, labor unions helped millions of workers win better pay, fringe benefits, and working conditions. They concentrated on the "hygienic" factors of the job.

Dr. Frederick Herzberg, now distinguished professor of management at the University of Utah, has segregated the factors relating to a job into two categories. Hygienic factors—things such as pay, working conditions, and interpersonal relationships—can cause dissatisfaction by not meeting a person's needs. Motivators—achievement, recognition, the work itself, advancement, and responsibility—determine whether a person wants to do a better job.[3] Although the hygienic factors have to be attended to, he says, they will not motivate people. Job enrichment attempts to redesign meaningless jobs so the motivators can come into play for the benefit of both worker and employer.

Some industry managers, and even government officials, have introduced job enrichment to their operations. Labor unions still concern themselves primarily with the hygienic factors and are among the strongest proponents for keeping jobs narrowly defined by restrictive work rules in order to artificially create or preserve jobs.

The Worker's New Expectations

The worker of today demands more from his job than pay. He wants a more fulfilling life. In the U.S. and other highly industrialized nations, he may ask for more time off rather than increases in pay. Or he may seek work that is more fulfilling.

Not everyone seems to be looking for a deep sense of satisfaction from the job. Some say they just want to earn a living or enough money so they can save up for a summer of travel. Yet, says Herzberg, they are just victims of a society that has taught them to deny their need to find fulfillment.

We have left the era in which people were driven by a sense of obligation to their traditions. This has significant impact on our attitude toward work. Herzberg observes that "the immigrant laborer was seeking security and a future for his children. He would bear his life on a Mickey Mouse job and gather his psychological income from his social traditions, ethnic traditions, religious traditions, political traditions, and family traditions."[4]

The immigrant laborer—whether from the farm to the city or from another country to this land of opportunity—raised children who now find these traditions empty in relation to their desire for "personal fulfillment and significance in what they do." They feel no obligation for work quality or productivity unless it relates to them in terms of their whole life.

When the U.S. was a survival society, people were concerned about goals. Jobs were a means, first, of surviving and, then, as a means for attaining one's goals—often for one's children. But, says Glasser in *The Identity Society,* increasingly, we strive for goals that reinforce our concept of ourselves as persons.[5] Marshall McLuhan believes the "whole pattern of life has shifted from job-holding to role-playing which means doing many jobs at once." A mother, for example, doesn't have a job; she has a role with, maybe, 60 jobs, he says.

Putting The Pieces Back Together

The search for one's role or identity places new demands on the workplace. The job has taken on new dimensions in our society. "The best place to start bringing meaning and substance back to people's lives is in the workplace since work is peculiar to the human species," says Herzberg. "Give workers the opportunity to grow through accomplishment and meaningful jobs," he advises.

Innovation will be needed to enrich jobs with enough content to permit the worker to find meaning and satisfaction in what he accomplishes. As for any dull, meaningless work that can't be made more humane, we can mechanize it. In a Swedish company which is a leader in trying to humanize the workplace, one may be surprised to see industrial robots performing some of the work. Why the incongruity? There's nothing incongruous about it. "We put robots into jobs where people were working like robots," explains a manager there.

Both enrichment and mechanization are likely to meet with opposition from some workers. People who have come to expect nothing more from their jobs than tasks that require little involvement are apt to resist any move that affects their job.

Job design cannot follow a single route because not all people are alike. We really have two different trends occurring, says Robert G. Frick, president of Case & Company Inc., a management consulting firm. In his contact with a cross section of businesses he sees "on the one hand, more aggressive, better educated, more independent young people, and, on the other, I see real problems with people coming out of the ghetto who are almost uneducated and very unmotivated."

There is a danger that, as we strive to make jobs more meaningful and challenging for motivated and educated persons, we will further close the door to those who are poorly suited for any kind of long-term employment. There are also many people in the middle who are not anxious to have their jobs changed. For example, a number of groups of U.S. workers have studied work-team assembly in Europe and a few U.S. plants. They generally have expressed a preference for their compartmentalized work over teaming up to assemble an entire product.

In recent years, a new element in the mix of people seeking employment has appeared—the "overeducated." Holders of college degrees, including doctorates, have not been able to put their education to full use and have settled for jobs as cab drivers, hotel night clerks, and blue collar workers in industry. Quite often, their specialty has little direct application in the job market even if jobs were plentiful.

The difficulties in matching people with work are part of a vicious cycle of social-educational-technological problems. We have to address ourselves to questions such as: Should our education system prepare people for work? If so, how can the education and business sectors do a better job of matching people with the available jobs? How can we prepare people to find more meaning in their lives both on the job and in their leisure time?

Designing Jobs For People

We can open the way to better lives by starting with improved job structure. But we cannot solve all our social problems in the plant or office. In fact, social problems originating outside the job limit what we can accomplish in the workplace. Even a good job cannot compensate for a worker's personal problems although he may direct his hostility toward his job, his fellow workers, and his employer. Donald Alstadt of Lord Corporation is disappointed by those behavioral scientists who claim that motivation problems can be solved entirely in the job. "The greatest factors in demotivating or motivating a person come from his personal life over which we have very little control, from his social life, from his exposure to society."

Humanization of the job will progress, however, for two seemingly different yet compatible reasons. It meets the needs of workers—to which a growing number of managers wants to respond. It also may be the only means of maintaining or raising productivity levels in a humanizing society.

Although productivity may be the prime motivation for some organizations to look into humanization efforts, that consideration alone cannot carry them to success. Some companies that have attempted to implement humanization programs have failed to make a go of it. They have tended to the hygienic factors such as working conditions and fringe benefits but have neglected to humanize the content of the job because that requires an understanding of people's inner needs and a dedication to working patiently with them to design meaningful work.

"If we are worried about humanizing organizations, we worry about it as a worthy end in itself and not merely as a means to some other unstated end," warns Roger D'Aprix, manager of employee communications for Xerox Corporation. He says there are more points where the needs of individuals and organizations correspond than there are points of conflict, however. In his book *In Search of a Corporate Soul*, he writes: "The person who argues that if we humanize organizations, we will produce groups of lotus-eaters who forget why they were hired is as incorrect as the person who claims that humanization will lead to great improvement in morale and, therefore, great increases in productivity."[6]

Productivity benefits cannot be guaranteed in a humanization effort, but prospects for them look far better than in authoritarian or paternalistic organizations if one considers the pressures in society. In this era of high education and high expectations, there seems to be little choice but to build content back into jobs, reversing the process employed in the earlier industrial age.

New York's Citibank has reduced its labor costs, upgraded job content, and extended more personalized service to customers by setting up minicomputer workstations for financial transaction services. Prior to the change, workers each handled just a piece of the transaction. The new strategy combines job enrichment with new technology. In a single workstation, a more professional job has been created in which the employee performs a range of functions to handle a complete transaction.[7]

We are well into an information revolution and now can see that the impact on jobs may be more beneficial than expected—not only in numbers but in content. The computer will not wipe out the middle manager's job, for example, as some had feared. It will give him information faster than ever, including some he has not been able to obtain in the past. This will give him better inputs for decision making.

J. Thomas Brown Jr., vice-president of the consulting firm Case & Company Inc., believes improved accessibility of information will enable the middle manager and other employees to enlarge their jobs. "On the one hand, it's going to eliminate some mundane calculations, but on the other hand, it's going to impose some other disciplines on him so his job is actually going to be tougher."[8] When a manager doesn't have the facts, his ability to manage is limited. But improved information means that his job will be enlarged.

Because business-technical-operational information will be more easily attained, the middle manager will be able to concentrate on the people factors in his job. Tending to human needs will confront managers with an infinite amount of work. The increasing proportion of "knowledge workers" in industry will compound that challenge because these people need a lot of human contact and demand a lot more information relating to their work.

Decentralization and the ability to call up data from locations through an organization will disperse decision making responsibility. "The average person—particularly in white collar jobs and service industries—is going to be more of a manager," says Dr. Neil Drobny, manager of the resource management and economic analysis department at Battelle Memorial Laboratories. "As classically thought about," he says, "the dividing line (between managers and workers) will fuzz up."[9]

Machines For People

Ironically, our society now sees the worker in more than economic terms just when we have reached a point in economic development where workers are more plentiful than capital. Capital for investment in innovation and production is in short supply, but people are not. Hazel Henderson, codirector of the Princeton Center for Alternative Futures Inc., believes we may have overshot the mark in substituting capital for labor. She contends that "in hundreds of production and service processes, labor has now become the more efficient factor." However, the question of whether labor is more economical than capital is a meaningless one, Mrs. Henderson believes. "We must ask, in all cases, 'efficient for whom?' " She suggests we consider the effect of any action on the individual, the corporation, society, and the ecosystem.[10]

These effects were not considered in the Industrial Revolution. Our personal expectations and technological capabilities have brought us to a place where we can create better jobs—a step toward meaningful lives. Since the economics have also changed, we now have to concern ourselves with giving people involvement in their work. "Our new identity society" has to find the means for transforming the jobs that were geared for survival—not involvement, says Glasser. And he reminds us: "If we had not already made a good beginning, particularly in plentiful good-paying jobs, we would not even be entering the new society."[11]

Technology has provided a strong economic base from which to work for a better society. It has even opened the way for changing the economic base itself. Whether we entrust a job to man, a machine, or to a man-machine team, the human factor—rather than machine design capability or old-style efficiency—will increasingly dictate the content of the job.

PART III
The Worsening Climate For Innovation

10 Overreaction by Overregulation

"To the frustrated, freedom from responsibility is more attractive than freedom from restraint."

Eric Hoffer
The True Believer

Racing to "new frontiers" and a "great society," the economy boomed. We could have anything we wanted. Yet, we could not extricate ourselves from a war we hadn't sanctioned and millions of Americans waited for their civil rights to be hammered out in the courts.

It was the 1960s—a time of social upheaval. Despite economic growth and the proliferation of technological innovations, the mood of the country darkened. We entered a period of severe self-criticism. We had no national goals, no long- or short-term objectives, no means of directing our efforts to the problems that were being exposed almost daily.

Unrestricted "progress" hadn't given us what we wanted so we placed limitations on one another—"here's what you cannot do." The can-do nation resorted to can't-do.

Most of the demands were laudable. But we did not ask what the impact of the legislation and regulation we requested would be on our basic system for delivering solutions.

"Historically, the American business system and the educational system . . . have automatically accepted incentives as the driving force for generating jobs, goods, and what we used to call a better way of life," says Edward N. Ney, chairman of the advertising agency Young & Rubicam Inc. In the 1960s, however, we began to reject incentives in favor of regulations as a way of getting things done.

Neither incentives nor regulations are inherently good or bad, says Mr. Ney. The question is how well each works in bringing the maximum benefits to the most people. "Incentive and regulation are not mutually exclusive concepts, but they differ importantly in that one is essentially a stimulant and the other a restraint."

Lost in the moral struggles of the 1960s and 1970s was the question of how to reach the objectives we had in mind. Coupled with negative feelings or neutralism toward technological innovation, our drive to regulate performance dampened initiative and creativity—our traditional means to better solutions. "Indelibly, our record shows that what powered Jefferson, Franklin, Fulton, Edison, Bell, the Wright brothers, Lindbergh, and Ford was initiative inspired rather than obedience required," says Ney.

"Is regulation a good means of reaching the objectives we share?" asks Dr. D. Quinn Mills, professor of business administration, Harvard Business School. He believes regulation encourages simplistic thinking about complex problems, sets up adversary relationships, and establishes rules that are rarely abandoned after they serve their purpose.

Regulations can serve a purpose. "National goals in such areas as environmental protection are necessary and—because of competitive realities—federal regulations are required for their uniform achievement," says Irving Shapiro of du Pont. "But," he adds, "we should not lose sight of the advantages in permitting freedom of choice in (selecting) technology to meet the goals that are levied on everyone."

The first thing to check when deciding whether a regulation is needed is whether there is any other way to handle the situation, says General Motors' environmental vice-president David Potter. Unfortunately, however, "at the moment, the judgment on this is in the hands of a bureaucracy paid to regulate. It's not likely we'll get an unbiased decision."

Conflicting Rules

Businessmen are not unconcerned with a clean environment or safe products or safe workplaces; however, they cannot correct all the ills overnight. Even if a company could get each of its competitors to follow its lead in, say, pollution control, it would not have sufficient funds to eliminate all pollution. Neither the funds nor the technological answers are immediately available.

If there were positive incentives built into our system of business, companies would respond with initiative. But they are forced to serve owners' demands for profits and the public's demand for ecological and safety considerations. Any of us might simultaneously be an owner, an employee, and a customer as well as a member of the public and impose conflicting demands on a company.

Industry, then, is forced to play a defensive game. And it doesn't know what the rules are. "Regulations are often contradictory, obsolete, difficult to administer, and impossible to prove that you have met," says Paul Chenea of GM's research laboratories. "With so many federal, state, and local agencies setting regulations, it's difficult to rationalize everybody's rules into one set you can follow," explains George E. Newman Jr., assistant treasurer, Hewlett Packard Company. A Denver Research Institute study shows that regulations are a significant obstacle to innovation. The main reason for their negative impact, however, is their "uncertainty rather than their stringency."[1]

One of the most frustrating aspects of regulation to the business manager is that of witnessing one regulatory agency working at cross purposes with another. An example: efforts to conserve energy which conflict with pollution standards. Another: antitrust regulations that conflict with efforts to boost employment and exports. And another: legislation to insure safety from toxic chemicals that imposes a heavy burden on small business when other agencies are trying to foster competition and promote small ventures that bring us needed innovations.

Managers see agencies causing trouble for one another and saying "that's not my problem." So the unclaimed problem creates uncertainty for industry, raising its costs, and forcing counterproductive efforts.

Who Regulates The Regulators?

The first National Forum on Business, Government, and the Public Interest, held in Washington in late 1976, drew leaders from all sectors. Discussion for the three days centered heavily on the regulatory situation. People from all sectors agreed that the U.S. badly needs some deregulation. Regulation has reached the stage of *overregulation*.

"We start off with a goal of safety, . . . But we bog ourselves down in design standards . . . in trying to define, in ever more detail, the idea of safety," said Harry Holiday, president of Armco Steel Corporation. "The same confusion has turned our search for cleaner air and water into a rat race of specified engineering demands.

"It's the obsession with methodology that's bogging us down. Regulators aren't being allowed to look for results. Instead, they are forced to blindly demand obedience to arbitrary, step-by-step rules. Innovative ways to reach our goals are stymied by our national obsession with bookkeeping."

S. John Byington, chairman of the Consumer Product Safety Commission, charged that "frequently, neither an agency's legislative mandate, nor the agency itself, articulates its real goals in precise, measurable terms. And, all too often, goals are confused with the means of achieving those goals." His commission, he said, is directed by law to protect the public against unreasonable risks associated with consumer products. The measure of success, he believes, should not be the number of standards promulgated nor the number of prosecutions obtained but the number of injuries and deaths prevented. Public administrators should be accountable both for the degree of success they have attained in meeting their basic goal and for the cost involved.

Cornell C. Maier, president of Kaiser Aluminum & Chemical Corporation, said, "the regulatory agencies have become a branch of government with few real checks or balances; but, more importantly, a branch of government with no liability for their errors. Some seem to think that the role of a regulatory agency is to protect the public from the excesses of business. But who or what is there to protect the public from the excesses of government? Who regulates the regulators? The power of regulatory agencies to stop the economy in its tracks, to maim or dismember whole industries, is staggering."

Lower and middle level bureaucrats "wield enormous power," agreed James C. Miller III, assistant director of the Council on Wage & Price Stability. Furthermore, they "believe in" regulation. The system should be balanced with some people who don't believe in regulation, he suggested. "Rather than reward bureaucrats for the number of fines they impose and how strongly they write regulations, reward them for exhibiting justice and moderation. Make it easier for government executives to hire and fire, and hold such executives responsible for the performance of their employees."

Mr. Miller also would make it mandatory for regulatory agencies to perform economic analyses of proposed regulations. Imperfect as these analyses might be, they could eliminate poorly conceived proposals, he says.

"Whether we call it 'regulatory reform' or something else, I believe we would all agree that a major reorientation of our governmental programs, policies, and attitudes is necessary if government is to meet the needs and demands of today's society," said Mr. Byington, reflecting the feelings of persons from both industry and government. "To achieve this goal, social programs must be specifically designed to require that quality-of-life goals and objectives be spelled out. Government administrators must be held responsible for answering questions of accountability: How many injuries have been prevented? How much cleaner is the air or water? What price reductions have been realized? Until such social objectives are required to be spelled out in planning and budget documents—a framework of socioeconomic trade-offs that make sense—regulatory reform won't be worth a damn!"

$2,000 Per Family For Regulations

Nobody is going to turn back the clock and eliminate our regulatory agencies, said E. S. Donnell, chairman of Montgomery Ward & Company Inc. "But many of us do think the continuous escalation of regulations is resulting in an 'overkill' that is dangerously counterproductive and economically harmful to the consumer."

Regulations cost Americans $130 billion a year, according to government estimates. That figures out to $2,000 per family. The Center for the Study of American Business has analyzed a partial list of the regulations in force and computed their cost at $65 billion.[2] Regulatory agencies spent $3 billion to administer these regulations which cost industry $62 billion for compliance. The largest entry was for energy and environment regulation at $8.4 billion. Consumer safety and health cost $6.6 billion; job safety, $4.5 billion. The center's costs figure out to $307 per American for this group of regulations. The study also concluded that the price of a typical 1976 auto reflected $557 worth of changes made mandatory from 1968 to 1976.

If people were aware of the fact that they are each paying several hundred dollars a year for regulatory protection, they might insist on a more prudent approach to regulation. They might not be able to agree on which goals are worth seeking in terms of the cost involved, but they would demand that overlapping and conflicting regulations be pruned away. And they would demand a careful accounting to see that they are getting their money's worth.

Too Much Of A Good Thing

The weight of the regulatory situation thwarts attempts to reach the worthwhile goals buried beneath it. Businessmen object to the excesses—not the basic concept—of regulation. They themselves have been responsible for the establishment of many regulations when they have sought protection or special concessions. They are aware, too, that regulations such as those pertaining to pollution control have paved the way for growing industries to deliver the equipment and solutions for compliance.

Responsible businessmen realize that we all have taken advantage of "free" air and water. But they know that obedience to regulations will not match the improvements that could come from more creative effort. And they fear that even obedience is impossible as the mountain of regulations becomes higher and more complex.

Zeal to reach quick solutions can actually increase our risk. Some of the regulations imposed on the auto industry, for example, have resulted in what amounts to experimenting on the public, says Dr. Craig Marks of General Motors' engineering staff. "Traditionally, we took an idea, put it on one line of cars and watched and developed it. With regulations, you decree that, as of a certain model year, you're going to do it across the whole model line. The magnitude of risk goes up."

Regulations have mushroomed in an effort to correct not just current wrongs but wrongs perpetrated for centuries. "We have gotten into the position of cleaning up what we inherited," says H. Ridgely Bullock, chairman and president, UMC Industries Inc. In addition, he says, we want to solve problems before we have the product that causes the problem. If the U.S. had put a supersonic transport into production, people would have worked on solving the problems it might have created. "What would have happened if, before we came out with the first auto, we had said it must be quieter than a horse?" he wonders.

Killing Creativity

The U.S. can stand the loss of a specific technological innovation, but regulatory zeal could seriously diminish our creative capacity. "We must not kick innovation to death with mandatory behavior patterns and standards," says Harry Holiday. "Human inventiveness needs room to operate. It thrives when it's encouraged. You can never order up ingenuity. Its appearance is random. But you can certainly stop it in its tracks."

Industry leaders realize that they must get this message across to the public. Their voices are too few, however, and their message is generally interpreted as narrow-minded, self-serving opposition to the public interest.

The public should be concerned at all times as to whether the cumulative effect of its public servants is serving the public interest. If regulatory momentum leads to a one-sided battle to show who is in control, the public has lost control.

The cost of regulations to each of us is, then, something far beyond the several hundred dollars that we can identify. What is the slowdown in innovation costing us in price reductions that we are not getting, or medicines that aren't available, or improved forms of transportation that are less harmful to the environment and our pocketbooks? Can we hope for a turnaround when we are nurturing a generation of industry managers who have worked in nothing but a climate of overregulation?

How can we hope for better solutions to the problems we have laid out if we allow regulators to specify exactly how to solve a problem rather than stating the problem and then permitting industry and individuals to come forth with innovative answers? The solar heating industry, which really isn't an industry yet, is having difficulty getting into gear because of the lack of standards in design. But industry and government groups are developing standards. This will facilitate progress to a point. However, progress is then likely to slow because of the expected plethora of mandatory standards and specifications that will box in innovation,[3]

Engineers from other nations puzzle over the high degree of standardization employed in the U.S. Standardization has ensured sufficient uniformity and safety for goods to be mass produced and mass distributed. Although standards may restrain the pace of progress, they do permit the benefits of an innovation to be spread widely. If, however, they are imposed on a development too early, they close the door to further improvements. And, if they specify too tightly *how* something is to be designed, rather than *what* it must do, they bar the totally new approach to doing the job.

The delicate balance needed could be hammered out in continuing negotiation. Few regulations are being set by negotiation, however. Dr. Marks points out that the goals engineers and scientists are working toward were set by arbitrary political forces rather than through logical scientific reasoning. Certainly, change is in high demand as we struggle to comply with new regulations. This has been a boon to the typical engineer, says Marks. But, he cautions, "I can show you people who are extremely frustrated because they know that what they're working on doesn't mean anything. Many of the 'numbers' we're working to meet were poorly set—'cut emissions by 90 percent' or 'double gasoline mileage.' " Manufacturers and regulators are quibbling about a percentage point or two here and there while the whole basis was politically established and may be unjustified in cost or benefit.

Regulations are being designed and administered in an adversary relationship rather than one of collaboration. One manager says his engineering staff and government agency personnel find it increasingly difficult to work together to solve problems. The government employees are expected to be watchdogging industry, so both groups feel uncomfortable about being seen working as a team.

A team, working on a scientific or engineering problem, can use an adversary relationship to advantage if they stimulate and challenge one another to find the best solution for a common problem. But a good guy-bad guy relationship does not produce the same results. It becomes a two-team contest with one maximizing objectives and the other minimizing its cost of compliance. The public would be better served by a pooling of knowledge and opinion to negotiate the best standards within present limitations coupled with a constant effort to overcome these limitations and establish higher and higher standards.

The Zero-Hazard Myth

Neither government nor a free business system can bring us risk-free lives, and a rush to perfection can have a negative, long-range impact on the system which has to deliver the answers.

Pollution standards, for example, were set to purify lakes and rivers without regard for the degree of cleanliness needed for man or fish or what it had been in the natural state. It took a long time before the cost of meeting these standards showed up in terms of threatened jobs and the decline in companies' competitiveness. Now, belated attempts are being made to negotiate reasonable cost-benefit trade-offs.

In 1977, industry purchased $7.5 billion worth of air and water pollution equipment. About one dollar out of every six for plant and equipment spending in the primary metals and paper industries was earmarked for pollution control. Billions more were spent to operate the control equipment in place. It's important that these dollars buy incremental benefits of greater value than the increments of expansion and innovation the companies sacrificed.

Dangers Of Delay

A major negative effect of regulation has been the delayed introduction of new products, says Union Carbide's Dr. Miller. "What took three years a decade ago now takes ten years. The increase in dollars is even greater because of the expensive testing."

The impact doesn't stop with "the company." It affects workers and customers. One of the most dramatic extensions in cost and time is in the development of ethical drugs. "We are faced with an alarming decline in the output of new drugs in the U.S.," says Dr. Gerald D. Laubach, president of Pfizer Inc. He notes that it takes about seven years and $15 million to bring a new drug from the laboratory to the practicing physician, compared with two years and $1.2 million in the early sixties. The number of new drugs approved by the Food & Drug Administration dropped by about 70 percent between the 1957–1961 period and the 1972–1975 span, notes Monsanto's John Hanley.

A new asthma drug to replace previous drugs that had adverse side effects was widely used in Britain for several years before the FDA approved it, says Murray Weidenbaum of the Center for the Study of American Business. "The delaying procedures not only increased business costs but, for an extended period of time, prevented American consumers from having access to the newer and better product."

Dr. Weidenbaum points out that the U.S. was the 30th country to approve the anti-asthma drug metaproterenol, the 32nd to approve the anti-cancer drug adriamycin, the 51st to approve rifampin to fight tuberculosis, the 64th to approve the anti-allergenic drug cromolyn, and the 106th to approve the antibacterial drug co-trimoxazole.

George S. Dominguez, director of government relations for safety, health, and ecology at CIBA-GEIGY Corporation, is concerned about the hidden cost and risk caused by the public's not having certain drugs. Doctors aren't able to do their best, he says. In Europe, drugs are on the market well before they are approved in the U.S. Because of this, there are signs that U.S. firms may step up their research and marketing overseas.

Innovators Ask "Why Bother?"

"One of the biggest deterrents related to technological innovation is product liability," says Ridgely Bullock. "If there is substantial risk on a new product you won't introduce it." On the other hand, there have been court cases in which companies have made a product improvement but were penalized for not having made it sooner, he continues.

"Political regulation tends to make positive planning very difficult," says W. R. Grace's Robert Coquillette. "It leads to negative planning—'what do we want to get out of?' " There are so many hurdles to putting a new product on the market, people figure "Why bother?" adds Neil Drobny of Battelle.

Complying with regulations, keeping up with the reporting load, and trudging through the bookkeeping is frustrating the people who should be innovating. "The crucial question," says Professor Carl H. Madden of the American University's School of Business Administration, "is whether moves to hold the corporation to high standards of conduct, to greater participation, disclosure, environmental concern and the like will or will not generate excessive burdens of regulation and control that stultify its productivity and that destroy the remarkably sensitive market system. It is, after all, the market system that has made our economy so productive and responsive."[4]

Weidenbaum urges business to warn the public of the adverse effects of overregulation but in a realistic fashion. "Strive for a meeting of minds with the public on what regulatory activities are actually excessive, but keep in mind that the public will regard others worth the price no matter what the impact on productivity."

The major corporations are in for more regulation, says J. Robert Harman, senior vice president of the large consulting firm Booz, Allen & Hamilton. "It's only a question of what form it will take." Some businessmen are hopeful about the future of regulation. They hope, for example, that people in the regulatory agencies won't be simply bureaucrats or politicians. Others hope that as business becomes more socially responsible there will be less confrontation.

Frustrating The Innovative Enterprise System

American business has been steadily concentrating into a few hundred companies which account for a high percentage of sales, profits, and employment. It is difficult to see a reversal of that trend. Some U.S. firms have to compete with the giant companies abroad—many of which are government owned or supported. It seems essential that the U.S. have some large-scale counterparts. Large companies also are the favored places of employment for the growing number of people who demand high pay, good fringe benefits, educational opportunities, and mobility.

Despite these pressures for largeness, Americans are highly sensitive to anything that looks like a threat to competition. Our antitrust mechanism, unlike that of any other industrialized nation, constantly watches for the bigness that in itself, we fear, signals an end to competition.

"The patent system is a stimulus to invention," says Xerox's Jacob Goldman. "Unfortunately, the concept of patents has been flying in the face of the interpretation given by the Justice Department and the Federal

Trade Commission on what constitutes monopoly. The patent, by definition, gives the inventor a monopoly. But the Justice Department and the FTC turn around and say 'if you're successful at it, beware.' "

W. Allen Wallis, chancellor of the University of Rochester, points out that innovation is a means of breaking monopolies. "Various monopolistic elements develop through government regulations, patents, natural monopolies, collusion or coercion, and these monopolistic situations, to the extent that they are successful in generating extra profits, are attractive targets for inventors and innovators. Efforts are made to develop new technology that will circumvent the restrictions by which the monopoly is preserved, and the greater the monopolistic profit, the greater the effort to circumvent the monopoly."[5]

The concept of bigness or smallness in business is misunderstood, says Donald Alstadt, speaking as president of a relatively small company in the industrial products and chemical fields. "I think there are situations where monopoly is necessary, and there are situations where it is deadly.

"I would rather call this country 'the responsibility and innovative enterprise system'—not the free enterprise system," says Alstadt. He attributes this nation's great success in the computer industry, for example, to IBM's domination of the field so it could earn enough profit to keep the process of innovation going. "Many parts of the chemical industry, on the other hand, have collapsed because of price competition," he recalls. "So competition isn't the golden symbol of our economy. Innovation is."

Our concern, then, should be focused on the climate for innovation. A company's competitiveness depends, among other things, on the state of the art in its industry and the possibility of competition from unexpected quarters. The airplane was not developed by the makers of autos, trains, or ships. The Polaroid camera was developed outside the photographic equipment industry. Synthetic fibers came from the chemical—not the textile—industry. The Xerox copier was invented by a lone operator whose idea was rejected by companies in the photography and office equipment industries.

Large companies have the ability to organize massive amounts of capital and thousands of people to bring a product to market as well as to engage in research. Small companies have proved to be a prolific breeding ground for new ideas. And the lone operator makes his contribution, breaking into established business circles with a good idea. Our regulatory structure is one of the most critical determinants of how effectively each of these contributors can play its role in the innovation system.

The way we manage our innovative process troubles Herb Hollomon of the Center for Policy Alternatives: "We have been mesmerized in this country since 1944 or 1945 that all one had to do in order to develop a new product or process . . . was to do research and development. As a consequence, when people passed regulations or changed tax laws that affected the innovative entrepreneurial spirit in the country, it was unrecognized."

Bringing new technology to market is a complicated process, he says, "and we have not yet attended in this country to an understanding of what governments do to the innovative process and how all these laws and regulations, incidentally, by mistake or through good intent, deny us the capability of bringing new things to the people of the country."[6]

11 Flickering Corporate Spirit

". . . we must move from . . . the primacy of technology toward considerations of social justice and equity, from the dictates of organizational convenience toward the aspirations of self-realization and learning, from authoritarianism and dogmatism toward more participation, from uniformity and centralization toward diversity and pluralism, from the concept of work as hard and unavoidable, from life as nasty, brutish, and short toward work as purpose and self-fulfillment, a recognition of leisure as a valid activity in itself."

Warren Bennis
The Unconscious Conspiracy

The innovation effort which blossomed into the new technologies of the fifties and sixties and provided momentum for new products in the seventies is a thing of the past. The climate inside and outside industry doesn't support technological innovation as it once did.

Despite the downturn in the statistical measures of our innovation effort (see Chapter 2) one might hope that the fires of innovation are still glowing in industry and that an upturn could come at any moment. Discussions with managers, scientists, and engineers in industry lead to suspicion, however, that the spirit of "the good old days" is burning out. The regulatory situation, social trends, and internal corporate stumbling blocks are dampening the incentive to innovate.

Although industry has made substantial gains in mechanical and management know-how, it applies them in an increasingly defensive manner. For a number of reasons, innovation generally carries a higher risk today than it did a decade or two ago, so gains in capability have been offset by the growth of conservatism. There are exceptions, of course. Some companies live by innovation. But far more have settled for new definitions of what it means to innovate.

Advancing Complexity Means Longer Lead Times

Americans feel that technological innovations are rushing at them with ever-increasing speed, but William S. Sneath, chairman of Union Carbide, says " . . . the complexity of our businesses and of our societies actually is advancing faster than technology. The time elapsing between conception of an idea and its ultimate fruition is steadily increasing. We must begin work on a new plant four or more years before the time we need its output. And we must begin work on a new product or a new process perhaps ten years or more before we will be ready to introduce it."

Many experts have observed that the time required to get a new product on the market has been shortening. Once an idea has been developed into a product, companies are good at bringing it into the distribution system quickly. In the earlier stages of development, however, the time required has stretched out. A basic innovation—not just a product modification—often involves complex technological exploration and a long period of testing for compliance with regulations on safety and pollution.

At one company, where bringing a product from idea to the store shelves takes an average of nine years, management confesses it is hard to be persuaded to put up money for innovation. And it's even harder to keep committed money coming, since at least one economic downturn can be expected to come along in that period and tempt management to cut off the funds.

This extended interval from idea to fruition runs counter to the pressures being put on business managers to score rapid-fire successes. Innovation is "affected adversely to a far greater degree than most of us think by the pressure put on management for a good showing, short-term," says Robert Gottschalk, a Chicago patent attorney and former U.S. Commissioner of Patents. "You can't do the things that ought to be done; you have to do the things that look good. This pressure certainly does a great deal to discourage the investment of corporate funds into basic research which, by definition, doesn't carry with it guarantees about when or how much the payoff will be. It also tends to discourage the grants that go to support the universities. And it tends to discourage the investment of substantial funds in even the more routine and predictably useful applications of technology," he says.

Jacob Goldman at Xerox explains how a businessman might choose to play it safe: "High interest rates and the need to pay off new things more quickly make the investor or the manager sit and think how he can better apply his dollars to realize the growth in earnings that he requires. He says, 'If I need a near-term gain, then rather than gamble on the 1-in-20 odds that one of my guys in the research lab is going to come out with an exciting new invention, I'll choose the better odds that I can improve my costs or add variety to my present line of products.' "

There is too much laying back and waiting to see what the other guy is doing, says Dr. Jeffrey Susbauer, associate professor of management, Cleveland State University. "Companies tend to go after the low-risk, low-return rather than trying to reduce the risk part of the high-risk, high-return opportunities."

Society, not just industry, suffers when too many people play it safe. Traditionally, industry has been the problem-solving sector of our system. But William Norris of Control Data charges: "For too long, business has been preoccupied doing the things that are the most profitable and leaving the solutions to most of the major problems of society as the responsibility of government."

Short Sightedness

Managers are often so busy protecting their position that they don't take time to ask basic questions like: "What products should we be in? Do we have the marketing capability? Do we have the technology?" They just want to keep a good thing going, says A. William Bloom, program manager at SRI International. They're looking at profits today. For them, "planning" is just a budget exercise. It's typical, he finds, that companies appropriate research and development funds without properly determining objectives for the program.

R&D programs should be part of the company's strategic plan, says Ed Jacobs, manager of SRI's R&D management program. "Top management should sit down with the marketing, manufacturing, engineering, and R&D managers and come to a common understanding of the company's goals and objectives, its resources—markets, products, technologies, and experience." Then, he says, "the strategic problems of what R&D should be doing and what level of funding it should receive begin to fall into place."

The innovation process can be initiated at the technical research end, the marketing end, or manufacturing stages. Eventually, however, it must bring together a product and a market. Despite their great technological success, the Wright brothers worked hard here and in Europe, looking for customers for their airplane. Whitcomb Judson invented the zipper, patented it in 1893, and then found that it was too expensive to manufacture; Gideon Sundback worked on a redesign of the zipper for seven years before applying for a patent in 1914.[1] The market for a sure cancer cure, on the other hand, anxiously waits for someone to come along with the right product or procedure.

Edison considered the sale of a product the sign of an invention's success. He didn't waste his time working on an invention if he thought there was no market for it. "In order to earn a profit, you have to do something for the customer that represents a better solution than he's had in the past," says George Newman at Hewlett Packard. He says his firm has frequently resisted the temptation to do a "me-too" job when a market has already been created by another manufacturer.

A company that is not innovative in terms of new products can survive as a "me-too" producer, but even that strategy bears a risk. Often, it is in the business because it can offer a lower price either through willingness to settle for a small return or manufacturing know-how that gives it the advantage of lower costs. Newman feels that a simple price advantage would not be a good reason for his company's getting into a business because the most significant contribution to profits comes from new products. But a new product in itself does not justify getting into a new line of business, he warns. The product not only has to make a contribution for the customer, it has to fit into the company's whole support system of manufacturing, distribution, and service, he points out.

Innovation can come in any segment of the long innovation process from conception of a product through manufacturing to marketing and service. Industrial companies tend to innovate in the segment they know best. A few are masters of the entire process. Since top management in the relatively new electronics industry is largely technical, they concentrate on product development. Other firms in more established industries—particularly those that have outlived the original founders who may have built the business on a new product—are more apt to seek innovation in their marketing or manufacturing.

A new consideration has been forced into the process in recent years, however. An increasing percentage of the innovation effort has to be devoted to compliance with regulations. Manufacturers have had to modify products or add features that may not extend the basic contribution to the customer and seldom permit cost reduction. Dollars have been shifted from developing new products to protecting those already on the market. In addition, the complexity of regulations and the inability to anticipate future restrictions has raised the degree of risk on a new product.

Some companies, obviously, have benefited from the markets created by regulations. Their innovation efforts have found an outlet in developing pollution control equipment, worker safety devices, etc.

Redevotion To Quality Is Coming

A more significant demand for change in the application for technology will result from new economic and social forces now at work on manufacturers. In the years ahead, "there will be a redevotion to quality," says Max F. Sporer, national director of management consulting services for Touche Ross & Company. Regulation is playing some part in this trend but underlying that is the scarcity of raw materials and capital. "In the sixties, manufacturers had to meet every whim of the consumer; they had to have full product lines," Sporer says. Now, he believes, we'll see a refocusing of attention on the old 80-20 rule which states that 20 percent of a company's products or customers provide 80 percent of the profits.

Manufacturers will be looking for ways to minimize the number of their product lines and the materials consumed in production. They will try to maximize the quality and durability of their products and make changes for the sake of improvement—not just to create a difference from last year's model. The auto industry made a break from its practice of planned obsolescence in styling several years ago; styling changes now come in three-year cycles. Industry may, at least, find relief from the frenzy caused by the short product life cycles typical of the sixties and early seventies.

The trend to durability will either depend on or necessitate a change in consumer buying habits. Either way, it would constitute a revolution for a disposable-minded society in which pens and razors are given a working life measured in hours and even calculators and wristwatches are inexpensive enough that people throw them away rather than getting them repaired—if they can be repaired.

Manufacturers will be involved in this sweeping change: Whether they lead it or follow it depends on their ability to develop products for an era when resources are no longer plentiful and to lead the public to accepting them.

Eyes On The Horizon

Because of shifting public attitudes and governmental involvement in technology, industry is making some changes in its structure and strategy so that it can continue to articulate ideas and be the nation's problem solver.

Some companies have taken steps to pipe technological expertise into the top management ranks and even to the board of directors. Gould Inc., like a few other large innovative companies, has a scientific advisory board of outside experts who scan the horizon, studying new technologies, market potential, and social impact. At W. R. Grace & Company, one officer represents the technical point of view when the appropriations board considers proposals by the company's operating groups. Corporate strategy rests heavily on good technology judgment. Continental Group has a vice president responsible for monitoring technological developments, studying the impact of new technologies, overseeing technology transfer among corporate laboratories and divisions, and studying new ventures.

Top management awareness of the technology picture not only helps it lead the corporation but creates a better climate for innovation. At Union Carbide, the management committee gives direction for R&D support to the twenty strategic planning units. By providing strategic guidance, says vice-president Thomas Miller, "you are telling people, 'If you produce something of promise, the corporation would like to be there.' It's disheartening to be in a lab, working for years on a project, and then have the corporation say, 'That's not for us.' "

"Not Invented Here"

Some great innovative companies have been ruined by mismanagement, observes Dr. Colodny at the University of Pittsburgh. He has seen companies lose their innovation lead by insisting that they use only their own patents. The head of a large research organization says there are many cases in which a company, or even an entire industry, pursues the first line of progress in a technology when alternatives would be better. The lack of vision or the not-invented-here syndrome can turn success into suicide. Sometimes it is abetted or imposed by government regulation; certain technologies, for example, cannot be imported from the Soviet Union, says one scientist.

Even companies in a competitive situation may fail to use innovation as a weapon. Managers throughout other industries express concern that the automotive industry has traditionally chosen styling, rather than technical innovation, as the basis for competition. They assert that the auto's basic configuration can be traced right back to the first horseless carriages which took their design from horse-drawn carriages and the industry follows cookbook rules which sustain a "miniscule" ratio of payload to power. Significant product innovation in that industry may have to come from a new company or new industry, they warn.

Time and again innovations that revolutionized a field most came from outside its borders. Take, for example, the Xerox copier, the waterbed, the electronic watch, or numerical control of industrial machinery. The electronics industry is upsetting applecarts in all sorts of other industries. Electronics producers are displacing numerous conventional products with their own cheaper and better versions or providing components so other industries can take advantage of new capabilities. Electronics has replaced mechanical and electrical devices in autos, office equipment, machine control, and even sewing machines.

In general, however, there is too little diffusion of technology throughout industry, says Elmer B. Staats, Comptroller General of the U.S. "The most advanced manufacturing technology is concentrated primarily in large firms, such as those in aerospace and electronics. But this technology has not been widely diffused to our medium and small sized firms."

William Norris agrees that "there is relatively little transfer of technology across industries. Yet studies show that many major innovations are the result of applying the technology of one industry in another. The bottleneck in industry preventing more of this happening is the concern for maintaining an exclusive proprietary position. In the private sector, we find a dichotomy between 'over-protection of rights to technology' and 'responsible sharing of technology for economic and social good'; a new attitude of cooperation by private companies is needed."

At Continental Group, Bruce Merrifield spends much of his time looking across industry lines to follow technological developments. He is keeping an eye on fiber optics, for example, because they are likely to have a significant impact in the communications field and thereby affect the market for paper—a commodity which represents a large part of his company's business.

The capacity to innovate depends on openness from the research end out to the market, says Ridgely Bullock of UMC Industries. "If you work at marketing not only to find out what people need now but what markets will emerge, that will indicate what technology is called for. Suppliers, too, are only too happy to tell you what's out front."

Locked Into Obsolescence

Companies have to be on guard against getting locked into obsolescent technologies. At the same time, the public must consider the fate of companies—owners and employees—if they are pushed into obsolescence. If, through regulations, we were to force the auto industry to make radical changes that necessitated scrapping some of its existing plants, what would be the consequences? We cannot afford to say "that's General

Motors' problem or Ford's problem." It quickly becomes a national economic and social problem.

There are some major companies whose assets are deployed from mines to manufacturing facilities. As we turn more and more to recycling and depend less on mining virgin minerals from the earth, we might push the mining facilities into obsolescence. Efforts to conserve materials and energy may create job displacements and threaten the viability of companies.

A certain amount of regulation-triggered innovation, however, may be healthy for some companies. Regulations are pushing the auto industry into innovation on a scale that few, if any, suppliers can remember. Some of them report a healthy attitude in the industry as a result of "its excited state." One comments, "The industry's engineers used to sit around and say 'it can't be done.' Now, they are jumping in to work with us, saying 'we can do it.' "

As industries mature, the thrust of their R&D shifts from product development to process development. With the product developed and selling, they emphasize capacity expansion and cost reduction. This is the stage at which a company begins to look more closely at labor costs whereas the blossoming company generally expands jobs rapidly, providing challenging work and opportunities for advancement. As investment in assets for manufacturing and distributing a product line grow, the willingness to risk it all with either a new product or major changes in manufacturing techniques diminishes. The larger and more mature the company, the more likely its base of ownership has spread, too. This adds still more reason for caution and entices management toward a maintenance rather than developmental attitude.

The Buggy Whip Syndrome

The share of corporate budget ventured on research and development varies considerably by industry. R&D dollars have a significant place in the budgets of technology-intensive industries, ranging from a low of about 3 percent of net sales to a high of about 12 percent, says Richard Atkinson, National Science Foundation director. In other industries such as textiles, lumber, and primary metals, the rate falls below 1 percent, he says.

Companies in the electronics, communications systems, and chemical industries tend to spend a higher percentage of the sales dollar on research, says Jacob Goldman. Basic commodities producers and railroads, for example, spend a very small fraction of their sales dollar for research. "Are they 'low-technology' companies because they spend so little? Or do they spend so little because they are low-technology?" he asks.

Answering his own question, Dr. Goldman says, "It's all a question of how a company defines itself. Think of a buggy whip company that might have called itself an accelerator company. After all, a buggy whip is just an accelerator for a horse. A smart R&D type in an accelerator company would be looking all over to see what it is that he can help accelerate. 'Anything I can do for you, Mr. Ford?' he would have asked."

All too often, companies decide how much to spend on R&D by looking at what they spent last year or at the average spent by their industry, observes Ed Jacobs at SRI International. "This approach doesn't take advantage of R&D as a major flywheel propelling the company toward its goal; and, not surprisingly, often results in an under-investment in R&D." You can under-R&D and overmarket or vice versa, warns George Newman of Hewlett Packard. About 10 percent of the sales dollar is what his firm finds it can spend effectively. Another member of the electronics industry, Dr. Noyce of Intel, says his firm has found 10 percent about the right pace to stay in front. But, he says, it all depends on the age and size of the company and the state of the technology in its field.

It depends, too, on how far into the future a company is concerned with product development and what it perceives as its long-term role. William Ylvisaker, chairman of the board at Gould Inc., says, "We're looking at products we can take to market in three to seven years and make a profit." Texas Instruments designates some of its spending as "strategic funds" which have the potential for launching new ventures that can change the direction of its business. William Wendel, longtime president of Carborundum Company, has said that the goal of R&D is to develop new businesses—not simply new products. Even relatively small Lord Corporation spends a high percentage of its revenues on R&D, calling these funds "new product futures."

The extent to which a company supports R&D depends on a number of factors: competition within its industry and from other industries, government regulations that affects its products, the need to find substitute materials, the availability of funds and personnel.[2]

A word of caution is offered by William Bloom. A company can devote a high percentage of dollars to R&D and not be innovative. Mr. Coquillette at W. R. Grace says that more and more companies are questioning the productivity of the money they spend for R&D. He believes that many of them are getting better results despite cutbacks in funds.

One way to improve R&D effectiveness is to size up the potential demand first. Bloom says companies frequently have missed their targets by failing to make a realistic assessment of the market potential for an idea early enough in the game. He knows of one company that wasted $2.5 million developing a good product for which there wasn't a reasonable market. An early $30,000 investment in market research would have shown that.

For every dollar of research invested in a product that is rejected in the market, about $10 is lost in all. The typical innovation effort includes about $2 for engineering and designing the product and $5 to engineer the production facilities. Startup of manufacturing and marketing account for the rest.[3]

It isn't the dollars for research alone that are restricting industry's innovation effort. They are a small part of the big picture. The large dollar amounts come in when you try to commercialize the idea, says Paul Chenea. "We could do all the energy research this country needs for a few billion dollars, but it would take hundreds of billions to implement it."

The Glamour Is Off

Setting up large expenditures for R&D is "no longer acceptable business practice," believes Mr. Coquillette. The dollars for R&D are looked at much harder than they were twenty years ago.

Corporate attitudes toward research have changed substantially over those years. Technology, in the modern sense, was born in World War II, says Sherwood Fawcett of Battelle. Some companies were introduced to it through government work; others observed what could be done. Demand for R&D came from stockholders, he recalls. "Just build a lab, put good people in it, and everything will come out all right," people thought. Companies frequently couldn't relate this to their business, however. The labs were simply showpieces. We went through a period of years, too, when numerous companies changed their names to include a term that connoted "technology."

In the late fifties and early sixties, says Dr. Fawcett, there was an increasing feeling that this research had to relate to the business. Then companies started saying the research had to show a profit in so many years or they would pull out. But there was still a willingness to gamble. By the early seventies, aerospace activity was depressed and the Vietnam war was over. People were concerned we had had too much R&D. "Instead of opening up our horizons, we said we had gone too far. On top of that, the stock market peaked for innovation. The glamour went off," he says. "You could announce a cure for cancer and it wouldn't be reflected in the stock market today," he says half-jokingly. "It probably wouldn't be approved," he adds.

Tight Money

Taking a hard look at R&D expenditures means rejection for some innovations because they lack big enough dollar potential to justify management's attention in a large company. Good ideas are buried in the dungeons of corporations because the market potential isn't big enough in relation to the total corporate scheme, says Dr. Susbauer.

There have been cases where a person with a bright idea has spun off from a large company and founded his own business based on a development that was too small for his employer. But he, like the independent inventor, finds it tougher and tougher to line up the capital, production facilities, and marketing structure to turn an idea into a business. The complexities of doing business today deter the small firm and the new venture.

New ventures no longer have access to the capital they need to exploit innovation. The funds available for technically oriented companies to get started have been drastically reduced in recent years. In 1972, there were 418 underwritings for companies worth $5 million or less; three years later, the number dropped to four. Although the total amount of new money raised for all corporations increased sharply during this time, the small company offerings dropped from $918 million to $16 million.

Battelle Memorial Institute points out that "recent legislation, however, well intentioned, has added to the growing conservatism of investors by cutting incentives to take risks." Financial resources, critical to emerging technical companies, are becoming concentrated in large financial institutions. Pension fund assets, for example, are expected to account for over half of all equity capital by 1985, says Battelle. Pension fund managers have to be cautious so they prefer to invest in large, proven companies. The individual investor has also lost the incentive to take risks because of tax rules.[4]

Because of the poor investment climate it is nearly impossible to amass a fortune today, says Neil Drobny of Battelle. That means there are fewer people to fund new research foundations. The existing nonprofit institutions have to be cautious because there is no one to bail them out if they run in the red.

Less Room For Entrepreneurs

Technological innovation in the corporation depends on the management of a variety of resources, not the least of which is people. The inventive activity "involves the supply of inventive talent, engineering skills, and the ability to bring specialized knowledge to bear upon the solution of technical problems," says Rosenberg.[5] The adoption of an invention "is likely to turn upon the supply of managerial abilities, highly motivated entrepreneurs, business acumen in perceiving market opportunities, and organizational flexibility and effectiveness."

Dr. Susbauer, who has done considerable study on entrepreneurship, believes there has been no decline in the number of entrepreneurs or potential entrepreneurs but that the number of opportunities for them to get into harness is shrinking.

The business-governmental relationship has raised the threshold for entry into new businesses, and the corporate structure of established companies often does not provide room for internal entrepreneurs. "When the boss, the system, and the subordinates all have to be served, there's not much left for creativity," says Susbauer. Corporate regimentation can narrow a person and hamstring his entrepreneurial talents. People who earn an MBA degree, for example, have to start working in a narrow function and this beats the broad view out of them. Then, when they move up to greater responsibilities, they have to be broadened again, he says.

Creativity and entrepreneurship can be smothered by an organization which tends to tell people what they cannot do rather than pointing out corporate objectives and asking them what they can do. The decision-making process sometimes suffocates creativity when its design is too cumbersome or individuals put procedure before substance.

Managing innovation requires abilities beyond those of other management roles. Edwin Gee and Chaplin Tyler, two long-time employees and officers of du Pont, have written a book on the subject. In it, they list several weaknesses that can result in failure when it comes to managing innovation:

- Poor interpersonal relationships such as pettiness or looking down on subordinates.
- Looking for intellectual solutions to problems and overlooking the subjective, emotional factors.
- Too much concentration on discovery for its own sake or the lack of business and marketing orientation which can lead to unproductive efforts.
- Failure to give enough attention to setting goals, providing assessment, and conducting follow-up.[6]

Creative Management

Creativity has to be fostered throughout the company, says Ed Jacobs. "If R&D is coming up with brilliant prototypes all the time, an uncreative manufacturing department may not be able to make products from them, and a plodding marketing department may not be able to sell them. And, of course, a stodgy top management may not understand the need for such new products or be willing to pay for their development."

Innovation cannot pervade the entire company if people are overprotective of their individual domains. Unfortunately, however, corporate structures frequently set goals for departments and use measurements of performance that actually pit one department against another. When flexibility would help advance an innovation, a department has to decide whether to react creatively or go by the book. A product designer, for example, may need to consult with people in the production department; the production people want to respond, but, after all, they have a quota to make.

One of the greatest organizational pitfalls in managing innovation is the tendency to build a large team either because of the workings of a manager's ego or sincere attempts to display commitment. But quantity is no guarantee of success. "With technological development, the quality of the efforts made is particularly important," says Curt Nicolin, a leading Swedish industrialist. "The quality of efforts is often determined by one or two persons. The number of people on a development team is therefore a poor measure of its capability."[7] Dr. Goldman agrees: "A good leader at the top of a research operation doesn't care how many people he has.

He'd rather have one Nobel Prize winner than 1,027 hacks. He doesn't measure himself by the number of people working for him."

Robert Coquillette says that several years ago W. R. Grace drew up a list of the company's "seminal thinkers"—those persons capable of contributing the truly big ideas. Out of 60,000 employees, 27 made the list. He explains: There is a difference between the person with the seed idea for a major innovation and those people who work on sequential, technical problems of implementation.

Top management may find it difficult to stay on top of what the researchers are doing. "Sometimes," says Mr. Bullock, "staying on top means being dragged by the heels by the guys below. The chance of success is probably less when the initiative comes from the corporate office, however."

Managing creativity depends on the climate you provide people, adds James Young of GE. "Put them on new things and they'll come up with new ideas." But he has observed that, while there are "some innovative people to whom you can give unlimited money and they'll pick the right things to pursue, there are others who can't tell a good idea from a bad one. They're not geared to appraisal. Take all their ideas, throw out the bad ones, and assign the good ones to someone else."

Bullock points out one important element in an atmosphere conducive to innovation that organizations often fail to provide. "You have to convince people that to try and to fail is not a bad mark against them. If they know they're going to be criticized for failure, they'll quit trying."

Closer To The Customer

The creative effort in industry aims at solving customer problems; making a profit at it sustains the whole innovation process. Some managers advocate, therefore, that the creative people be exposed to both the customers and the company's financial performance. "If you want to keep research people motivated, you want to keep them close to the product so they can see the results of their work coming out into the end product and finally showing up on the profit-and-loss statement," says the head of one high-technology firm.

Most large technology-oriented companies conduct the bulk of their R&D at the divisional level, close to the market. Some even put product development in the hands of technical entrepreneurs so that they, in effect, become the core of the sales force. This orientation serves well for a major portion of the innovation process. But there is disagreement over whether it serves the researcher at the front end who is seeking new knowledge; perhaps he should not be concerned with end products.

Many of the larger companies have centralized research laboratories where more general research can be conducted than in their divisional, product-oriented labs. They can advance the state of the art and explore new technologies that may ultimately apply to the company's products or its manufacturing processes.

Some of these centralized laboratories have been shut down in recent years, however, and others have been diverted from fundamental research to short-term product development. This is reflected in the numbers that show industry's allocations for basic research have fallen behind inflation in the last decade. And they represent a smaller share of the R&D budget—about 4 percent in 1976 compared with 7 percent in 1966.

Proud as they are of their technological innovation efforts, most industry managers feel that basic research is not their business—that it should be done by government, the universities, or private research organizations. With the emphasis on product development and modification, an increasing percentage of the creative team, therefore, is being exposed to the action end of the business.

The Risk Of Risk Avoidance

Managing a company for innovation today challenges the chief executive to deal effectively with his experts in each aspect of the business. He must lead the study of proposed corporate actions at the interface with government, customers, employees, and stockholders. He has to provide the climate in which people can grow and find self-fulfillment.

"Managing technological development demands true qualities of leadership; it demands not only technical expertise and intuition, but also a great deal of courage since the risks involved are great," says Curt Nicolin.[8] The risks are greater than ever because managers have to find a course through economic, political, and social uncertainties, but willingness to take a business risk is vital to the innovation process. The job of management is to minimize the risk in a proposed course of action. If, through internal weaknesses and external forces, industry shifts from risk taking to risk avoidance, the innovative spirit will die out.

The greatest risk a manager faces today is the possibility of losing his own willingness to take the business risk on behalf of a good idea.

PART IV
The Democratization of Technology

12 The People Will Decide

"For democracy is a method of finding proximate solutions for insoluble problems."

Reinhold Niebuhr
The Children of Light and the Children of Darkness

Fears of a swine flu epidemic mounted throughout 1976. Massive quantities of vaccine were manufactured and rushed to the public. But most people turned down a free shot.

Technology had permitted us to anticipate a problem and attempt a solution. But people were skeptical of both the "problem" and the "solution." Some doubted that there was really a problem; others feared that the innoculations were useless or might cause adverse side effects. Within a year, suits were brought against the government on behalf of people who had allegedly become ill or died as a result of the shots.

In the autumn of 1977, there were warnings of polio and measles outbreaks in the U.S. Had some people taken the prevention of disease so much for granted that they wouldn't bother to have themselves or their children innoculated?

This may be the age of experts, but Americans have shown that they will not always take expert advice. A quarter century ago, James B. Conant warned, "Among the highly significant but dangerous results of the development of modern science is the fact that scientific experts now occupy a peculiarly exalted and isolated position." The unfortunate thing about technological controversy, he said, is that the "number of people qualified to take part in the controversy is highly limited."[1]

They may not understand the scientific facts, but people are injecting themselves into the controversy. They may realize something Conant, the great educator, took pains to point out: a scientist is not "cool, impartial, detached." The convictions and pride of authorship "burn as fiercely among scientists as among any creative workers."[2]

We have generally allowed a free socioeconomic system to bring forth technological innovations in a fairly random manner. On special occasions—war or the space race—government provided deliberate thrust along certain lines. In recent years, we have turned increasingly to Congress and the regulatory agencies to make decisions regarding technology.

But people are questioning whether our present institutions—governmental and private—are capable of sensing our needs and translating them into action based on long range, global considerations. Addressing the Limits to Growth 1975 conference, Sicco Mansholt, former president of the Common Market, said, "We don't have the machinery to act even if we could decide what we want to do." Short-term interests conflict with long-term necessities, said Mansholt. " . . . the guiding light for political action is to win the next elections, to stay in power and, even in autocratic systems and in Russia, it seems to be the same."

Futurists around the world agree essentially that the limits to growth are set by institutional limitations—not physical resources. They question whether our social and political systems can apply technology to promote rational growth and provide a better life on an equitable basis.

We face complex challenges and problems which require long-term sometimes painful courses of action. Can we expect them to be solved by politicians whose horizons are as close as two, four, or six years? It took years following the oil crisis of 1973–1974 for an energy policy to be proposed. Like any comprehensive policy, it did not present quick answers that would make everybody happy. Instead, it reflected the host of interrelated problems associated with the energy supply. It, therefore, fell victim to political surgeons who hacked at the limb or organ of their choice.

Controlling the effects of technology "is likely to become a prime challenge for the civilization of capitalism in the next generations," writes Robert Heilbroner. This problem is "rooted in the nature of industrial society—not just capitalism," he says.[3]

The Changing Political Process

Since unbridled technological advance and undifferentiated economic growth are unacceptable, someone has to make judgments as to where we shall proceed. But people express a lack of trust in experts, the business community, and government.

Americans believe in the right to make the choices affecting their lives. If the conventional mechanisms through which they delegate their decision making are becoming ineffective, they may call back the powers to themselves.

Marshall McLuhan declares that "politics are dead. There will be no more parties, no more policies, just charisma. TV is the end of politics; all you have left is the image," he says. "Representation belongs back to the litany of the horse and buggy period when it took a long time to get to 'headquarters.' " McLuhan seems to be right, considering the difficulties representatives and political parties have in trying to find a coherent pattern in the thousands of demands on hundreds of issues. Hopefully, we have more than charisma left, however. Eric Hoffer warns us that a society which depends on charisma is in danger.[4]

Charisma can set a tone. It can represent a few basic values. But it is limited in how much detail it can convert into sound action, and it offers no promise of expertise in dealing with a complex problematique. It may even point to a goal, but it cannot clear the path that is overlaid with scientific unknowns, economic trade-offs, and social bramble bush.

We're seeing the end of the nonideological party system, said Daniel Bell at the World Future Society general assembly in June 1975. "Issue politics is replacing party politics." We have an overloaded system because of rising "entitlements," he warned. There has been an important shift in the way economic and social changes come about, says Philip Lesly, a consultant in communications and public affairs. The center of action is the group of "Power Leaders," he says in his book *The People Factor*. This group includes Congress, the federal administration, the courts, regulatory bodies, and their state and local counterparts. " . . . the Power Leaders are the target of heavy input from the articulate opinion leaders—the media and educators—and from the newly emerged vocal activists such as Ralph Nader, the ecologists, advocates of welfarism, and others."[5]

The degree of attention that these aggressive inputs receive is out of proportion to the public's concern. Furthermore, says Lesly, "it also creates a far greater sensitivity to those expressions that are received from the general public than their seriousness warrants."

The power leaders, the opinion leaders, and the vocal activists "form nearly a closed loop of communications," Lesly charges. "The activists and the media need and use each other, and both exert far more impact on the Power Leaders than the vast mass of the public or any private institutions, including business."[6]

Newspaper headlines reflect the tone of the new form of influence. Generally, we are on the threshold of disaster—past the point for any corrective action short of a swift reversal of what we are doing. There's little, if any, discussion of long term objectives, complications, or evolutionary changes that might resolve the problem.

Society has problems—problems that are not merely the creation of opinion leaders or activist groups. But effective solutions have to be reached by setting priorities and balancing the effects of corrective action for one problem against another problem. The rush to simple answers is the most dangerous aspect of today's emotional atmosphere.

Lesly has more recently commented that the public continually demands "quick fixes" of massive problems. In addition, he says, "new ideas are adopted readily by millions—so long as they seem to promise easy answers and as long as they are not identified as the recommendations of persons or groups in authority. More than ever, ideas are acceptable to Americans in inverse ratio to their identification with power centers. Ideas are most likely to catch on if they come from somewhere within the populace itself, or seem to."

As the public's expectations have risen, so have its many voices. Direction of our technology is going to have to involve public participation—unions, consumers, environmentalists, businessmen, and others,—insists Egils Milbergs, a policy analyst at SRI International. "But, right now, we really don't have adequate processes for involving people in the development of policy."

Fletcher Byrom of Koppers Company sees technology as a powerful weapon in social discourse because of the fact that it has called "for more human participation in decision making, not less. It provided the instant and full communication that is the enemy of covert power." Under an authoritarian government, says Revel, "the facts are the only things that enjoy freedom of speech, and the people must wait for them to speak—that is, they must wait for a catastrophe—before being allowed to ask where they are going." A democratic government differs in that it permits people to foresee disaster and adopt new goals or means, he says.[7] "Free information allows a solution to be found before a catastrophe occurs," says Revel. "I do not mean to say that a free flow of information is always sufficient to prevent catastrophes; but it is obvious that it limits the number and extent of such events."[8]

Revel contends that the U.S. is the only country capable of leading a global revolution. In fact, the revolution has already begun, thanks to our economic prosperity, scientific and technological advancement, and our spirit of freedom.

Messy Mechanism

Although we might like some sort of neat system for expressing our values and wants, "we've already demonstrated that we have a mechanism," says Continental Group's Bruce Merrifield. "We arrive at a consensus after discussion, turbulence and debate." We shouldn't mistake the turbulence as an answer—or lack of an answer. Turbulence does not mean that consensus is impossible. He points to the Watergate issues where confusion and controversy finally resulted in the system's working to correct the situation.

Cumbersome as the system is, people react. Industry, the scientific community, and our representatives are not going to get specific guidance from the public on every issue, but the public does set the general course, say believers in the system we now have. "There is a great wide road you can travel down, and people won't bitch until you get to one edge or the other," says one corporate officer. People will speak up when an innovation comes along that they don't like. And, if one they want is blocked, they will demand change. "When we over-control in an area and enough harm is done, the electorate then calls for correction," says Robert Coquillette. "In a few years, when we have brownouts and can't put in new manufacturing plants because we don't have the electric power, ideas of what we can tolerate with respect to nuclear plants and coal-burning utilities will change," he suspects.

The American system may offer the best hope for meeting the needs of the long-term future, but that does not preclude attempts to improve it. We may not be able to neaten the mechanisms, but we can improve the dissemination of facts to the public and make the system more responsive to people's wants.

Alvin Toffler, in both *Future Shock* and *The Eco-Spasm Report,* has called for involving the public more actively in decision making. He advocates paying more attention to the distant future with widespread participation in the process—"anticipatory democracy." Without this, he warns, "even the most conscientious and expertly drawn plans are likely to blow up in our faces."[9]

Anticipatory democracy would include an elaborate web of information systems through which people could express their wishes and receive feedback from their institutions. It would continuously blend scientific fact with human values in the search for the best courses of action. It would rely on a host of media such as goal-setting groups—ranging from neighborhood to statewide, futures research and long-range studies by Congressional committees, audience feedback programs on two-way television, and the referendum.

Involving people in their political system is not simply a matter of ideology. It is a practical necessity. One of the most urgent social inventions we need, says Aurelio Peccei, is a means to transfer power to the people or provide them some role in resolving the critical issues.[10]

John W. Gardner takes a slightly different view of participation: "Contrary to fashionable views of the moment, I do not believe that the urge to participate in the shaping of one's social institutions is a powerful human motive. But we must fan that uncertain flame." This is a time when people are withdrawing their faith in institutions, he reminds us. "It is not essential that everyone participate. As a matter of fact, if everyone suddenly did, the society would fly apart." The important thing is that "participation should be an available option"—that people *can* participate if they want to.[11]

Gardner believes, as did E. F. Schumacher, that the best way to get people involved in their institutions is to humanize and localize these institutions. It is difficult to feel individual responsibility to a huge, distant government and when one cannot see the consequences of his actions. "People can be themselves only in small comprehensible groups. Therefore we must learn to think in terms of an articulated structure that can cope with a multiplicity of small-scale units," Schumacher asserted.[12]

Peccei's sense of urgency regarding popular participation comes from the conviction that billions of people are no longer willing to sit idle while others shape their lives for them. The masses have a new power even if it is nothing more than the power to disrupt systems which enslave them.[13]

Man has technological powers to refashion the planet and shape his own destiny, but he will live in crisis until he learns to bear this power with responsibility. Acquiring the ability to govern himself may be the next step in the evolution of man just as prolonged inability to do so may be the end of him. Without a sense of personal responsibility and participation, there cannot be much hope for attaining the good life. Peccei contends, "the present global crisis, in which everything in the human system seems to be out of balance with practically everything else, is a direct consequence of man's inability to rise to the level of understanding and responsibility demanded by his new power role in the world."[14]

Informing The Public

Giving direction to technology will necessitate the flow of scientific facts and information about human values and wants. Participation by an uninformed public would be no better than control by a scientific elite. The U.S. has the technological and political foundation for free communications. There is a flow of constantly improved hardware for communications because Americans want information. A society that has nurtured science and technology has become accustomed to insisting on the freedom to get the truth.

We may doubt whether the layman can understand scientific facts if they are presented to him and, therefore, whether the public can make sound decisions about the long-range future. The man in the street can understand the consequences of a technology, insists David Potter. "You can bring almost anything down to him. If this communication doesn't happen, that's the scientist's fault."

The scientist, the businessman, the economist, and other experts are obligated to play a part in the communications mix. They have access to certain facts that should be shared. But they are hesitant to speak up, and the public skeptically wonders if they are hearing all the facts or just the selected facts and opinions that serve the expert's interests.

"There is a reluctance within the scientific community to present complex data in the public forum," says Monsanto president John Hanley. "Yet, only through a careful explanation of complex issues will the general public begin to understand that all public issues involve risks and benefits."

The public becomes especially confused or skeptical when the experts don't agree with one another. It hears economic experts who can't settle on methods for creating employment or controlling inflation. It watches ecology experts disagree with one another or with other scientists. It follows the activities of self-appointed consumer advocates and wonders how they acquired the credentials to speak for everyone.

Controversy And Complexity

Even fundamental "truths" are open to question. When does life begin? When does life end? Through technology, we have been able not only to dramatically reduce the incidence of death at birth, we can prevent birth itself. We can do such a good job of sustaining life functions that we can prolong a sort of life after "death."

If the experts can't agree on when a person is alive, on whether a certain food additive in a given quantity causes cancer, on whether a nuclear facility is safe, how can the man in the street know?

The rush of information—opinions, certainties, and uncertainties—presents man his first lesson if he truly is to become the caretaker of the earth and of his own destiny. Simple and clear answers are for the irresponsible. The responsibility offered us comes loaded with complexities and interdependencies. What we face is not *the* truth but a bundle of truths. Only our values can give us direction. For this reason, the public—not the technocrat nor the politician—must make the decisions.

In the matter of values, the public is the expert. Collectively, we—not the physician—make the final determination, for example, on what life is and what death is. The physician makes observations for us, but we decide what those observations mean in light of our values.

Greater participation by the public invites the expression of conflicting values. This raises the level of controversy and complexity—the price of progress, freedom, and vitality. Perhaps, Victor Ferkiss writes, "images and style are what politics will be all about in a technological society. The computers will solve the simple problems, leaving only the gut issues for the citizenry."[15] We might like to reduce all our problems to mathematics, but our biggest problems are created by people's nonquantifiable feelings.

Sorting Facts And Feelings

Distortion of the facts with feelings has led us to today's shortages of materials and fuels. We harmed our environment through bad choices. On the other hand, we have stalled technological and economic progress in cases where we erroneously thought they would harm the environment.

Pressure groups generally attack through our feelings—not facts. Government agencies are peopled by people so they, too, are subject to feelings. The courts can't provide the best decision making because they lack the expertise in technical, economic, and social matters.

Eventually, we will have to arrive at a point where we all feel a responsibility for managing our lives and our environment. If and when such a time comes, we will face the "gut issues" with an approach that differs from today's in several ways:

1. We will live with uncertainty and learn to err on the side of caution while preserving a willingness to take worthwhile risks.
2. As new facts come to light, we will update practices, customs, and laws.
3. An informed public will be aware that it is weighing the risks and benefits of a proposed action against the risks and benefits of alternative actions. It will know, too, that even inaction is a choice.
4. We will find personal identity and involvement in small groups without limiting our concern to the local context. We will be aware of the global consequences of our actions and responsive to the changes that global developments urge on us.
5. We will separate scientific fact from feelings in order to define problems and propose solutions. We will then be led *to* those things that are most desired rather than *by* those that are simply possible.

If the public is to manage its technology, we will have to find ways for it to acquire the information it needs. Information moves relatively freely in the U.S., but the public does not always see a full, accurate picture of a situation. Because they are free, the information media may select the information or point of view that is most helpful to the interest groups closest to them or that makes a good story.

Some people in the scientific community advocate establishing a science court to deal with controversial technological issues. The court would help bring out all the facts concerning the merits, faults, costs, and consequences of a technological development. The court would not make final rulings. It would simply attempt to serve as a forum for bringing information to the public and its representatives. Arthur Kantrowitz, a leading proponent of such a court, believes that it would demonstrate to the public that the scientific process is an adversary process. It would also compensate for the scientific community's failure to communicate with the public. Kantrowitz's concept of a science court provides for debate among expert advocates from the field under discussion. Judges selected by the advocates would rule on the scientific evidence, and the court's findings would be published. Other science court advocates suggest conducting the hearings before Congress or on national television.

A science court might be of some help, Egils Milbergs allows. "But who are your experts going to be and how are you going to pick them?" he challenges. Ed David, the former presidential science adviser, supports the concept but is concerned that "the public, the legislators, and the policy people for whom the output is intended will get the idea that scientific truth is determined by a quasi-judicial process rather than work in the laboratory."[16] An even more dangerous assumption could arise. The public, bowing to the superior knowledge of the court participants, might assume that it should withdraw from the controversy. In time, the court could take on the mantle of final authority.

No Future Without Risk

Whatever the mechanism for studying our future moves, the element of risk will not be removed. We extend and apply our knowledge to minimize risk, but there always remains something unknown and unexpected. Man has always lived with risk. "The fact is," says George Bugliarello of the Polytechnic Institute of New York, "just by standing up we have risk, the risk increasing from the very fact that we're walking on two feet rather than four."

A couple of years ago, a television program on the pros and cons of nuclear power opened a segment with scenes of a nuclear power plant located near Plymouth Rock—symbols of our willingness to take risks. We sometimes forget that the past has had its risks. We fear the dangers of uranium mining, nuclear power plants, and radioactive waste disposal and forget that we have endured coal mine cave-ins, oil tanker spills, and refinery fires.

Democracy itself is a form of risk. We put up with the messy mechanisms we have because we know that no one has all the right answers. The successful democracy is not led by sheer numbers, however; it listens to the best that anyone has to offer. "Why submit to democratic decision making if one knows the ultimate Good and has a scientific theory that determines how a society must be managed?" Revel asks. "Democracy is based on uncertainty. If the people believe the leadership is in error, they can replace them with other leaders. The opinion of the majority determines the society's direction because no one is totally committed to belief in an unarguable Verity."[17]

Designing A Resilient System

We sometimes get the feeling that society is so complex that we can no longer trust that individual actions will add up to the benefit of all. More and more, we sense that we have to coordinate our activities. The actions of one company can jeopardize an industry. One industry can disrupt the economy and our attainment of social objectives. One product can endanger the entire human species.

This is raising questions as to how far the U.S. should move toward national planning or even a planned economy. But, ever protective of their right to make independent, free choices, people ask some tough questions. Who would do the planning? Politicians with an eye on the next election? Technocratic experts or economists who are no more expert than anyone else in dealing with value trade-offs? What nation with a planned economy can rival the U.S. in material gain, individual liberty, or responsible citizenship?

What we need is "alternatives analysis, not national planning," insists Fletcher Byrom. He has learned from business experience that one can predict what might happen but not what will happen. Studying the alternatives, he believes, can lead to alternative mechanisms which can be readied to respond to events as they unfold.

Canadian ecologist Dr. C. S. Holling says, "We are all the better for surprise. We have talked too much about risks being bad. We try to engineer out failure. But then we run the danger of constructing a nonresilient system."

The danger of planning in the political arena is that it will lead to rigidity rather than adaptability. A government executing a plan cannot allow failure because everyone would suffer. So it goes beyond forecasting and setting an agenda to controlling. When it tries to bar surprise, it must bar some freedoms. If it has to depend on coercion, it has to suppress creativity.

The Blessings Of Diversity

Somehow, we must strike an extremely delicate balance between safeguards against catastrophe, on the one side, and maintaining freedom and creativity to discover and solve problems, on the other. Somehow, too, we have to set priorities continually for applying our talents and resources because not all problems and opportunities are of equal weight.

The book *1984* created an extreme example of life without freedom and creativity. It portrays the excesses of power—not the excesses of technology. Although reference to the book so often brings to mind the two-way television screen used by Big Brother to pour out his propaganda and to monitor people's actions, this was not a high-technology society.

People were poor in both material goods and spiritual freedom. Cigarettes fell apart, chocolate was tasteless, and the coffee was bad. An underground book read by the central character tells how technology had raised living standards in the late nineteenth and early twentieth centuries until the widespread increase in wealth threatened the hierarchical society which was supported by poverty and ignorance. In time, science and technology fell short of expectations because the type of thought needed to support them could not survive in a regimented society.

Diversity, not regimentation, will be our salvation. The ecologist tells us that uniformity is not good—that nature did not structure the world on purity or unity. An ecological system has fluctuations; things deteriorate, die, renew themselves. We have learned that even a certain amount of destruction by fire is important to the survival, diversity, and richness of a forest. Our political-social structures, likewise, should provide for variability so that people with differing goals and values can live side by side.

"Living inexpensively, enjoying the many benefits of technology, doing our own thing—nobody laid that out or planned it," says former science adviser Ed David. "It's the result of lots of decisions and that has to be the way it goes. Any attempt by government or environmentalists or a business association to mastermind it would fail."

We do, however, share enough values with which to bind a society together. For example, we still value such things as justice, liberty, equality of opportunity, dignity of the individual, brotherhood, and responsibility, says John Gardner.[18]

Man In A New Light

The individual's day-to-day problems of survival have diminished in advanced nations but the survival of mankind has become more problematic. We are pessimistic about the future of man because the scholars of man have, as Glasser puts it, "erroneously concluded that aggression and antagonism, not cooperation, are innate human qualities."[19]

According to Glasser, man has lived through four basic societies. Until about half a million years ago, man lived in a primitive survival society characterized by intelligent cooperation in the battle against nature. Then, overcoming some of life's dangers, he was free to enjoy the society of his fellow man. With the dawn of civilization and agriculture about 10,000 years ago, he entered another survival society, struggling against "a hostile environment almost all of his own making." This is the man we know and about whom we are so pessimistic. But by about 1950, says Glasser, a civilized identity society dawned. Rather than being goal-oriented, man is now becoming role-oriented. Rather than competing for power, he seeks involvement and cooperation.

Although we have not eliminated aggression, we should allow for the possibility that man will evolve a more cooperative society. It is not man himself but his societal forms that need uplifting. Peccei's thesis is that "we are not really facing a crisis inherent in man's nature, an irremediable biopsychical human flaw, but a crisis of civilization or rather a crisis of cultures, entailing a profound incoherence of human thinking and behaviour in front of a changing real world."[20]

Daniel Bell observes that man's first concern was with nature and then with remaking nature. "The post-industrial society turns its back on both," he says, as men "live more and more outside nature, and less and less with machinery and things; they live with and encounter one another."[21]

In the U.S., adversary relationships have been prized; and "healthy competition," honored. But our encounters are becoming indirect; we discuss or argue less in person and more through attorneys. We have built a highly litigious society in which the legal profession has become the growth area for employment. A recent count showed that 291 of our congressmen and senators were lawyers. Washington is known as "the most lawyered city in the world."

Litigation gets in the way of cooperation and can suppress freedoms even as it tries to protect them. Despite its technological strength, the U.S. may have to learn something of cooperation from other nations. In Japan, for instance, harmony is a fundamental principle whereas Americans see themselves as something apart from the group, observes one Japanese businessman who has worked in the U.S. Our alternative is to invite more authoritative rule. Until we mature as a society we may have to curtail personal liberties.

Although he is an optimist, at least concerning the long-range future, Fletcher Byrom warns that we are going to be less free in the years ahead because our culture is failing to deal with the technological questions raised. Inability to deal politically and socially with the energy problem, for example, is going to lead to some kind of authoritarianism, he says.

There are those who fear that technology inherently leads to centralization and the loss of freedom. But appropriate technology is equally capable of permitting us to create a society that enjoys flexibility and freedom. "The problem is not whether man can survive regimentation and standardization," Toffler says. "The problem . . . is whether he can survive freedom."[22]

There is also the question of why we want freedom. It is not an end in itself; the purposes for which we use it determine whether it is a value worth striving for. Werner Dannhauser, associate professor of government at Cornell University, says we have to distinguish whether we are interested in "freedom from" or "freedom for." The latter implies that we will use freedom for the sake of something higher.[23]

The freedom needed for man's survival and evolution is the freedom for seeking knowledge, influencing what's going on in the world, and becoming a total human being. With this kind of freedom, we could look to the future with the expectation that dreams might come true because we would be fashioning the future. We would not apologize for what we have come to know or the mastery we have earned.

13 Private Enterprise and Public Goals

"The problem that we face is a collective responsibility in order to balance on the one hand the competitive enterprise system from which innovation, entrepreneurship and new products and new processes come and, on the other hand, a sensible series of restraints. To try to wish away one or the other, to say that government can produce the products and services, or that the enterprise system can take care of the public good, I think, is misreading the times."

J. Herbert Hollomon
Director, Center for Policy Alternatives

This has been the land of the entrepreneur. Americans have been masters at translating ideas into production processes and products that have resulted in better incomes, better jobs, better health, and greater security.

Unfortunately, most Americans do not understand the enterprise system that made all this possible nor technology which the system employed. But even Karl Marx recognized the fact that capitalism brought unprecedented increases in productivity and mastery over nature. Rosenberg observes that Marx and Engels knew that "unlike all earlier ruling classes whose economic interests were indissolubly linked to the maintenance of the status quo, the very essence of bourgeois rule is technological dynamism."[1]

The profit system rewards the innovator. Not everyone with an idea for a new product or for getting a product to market is richly rewarded. But the possibility of financial gain and the opportunity for personal satisfaction drive our system with a continual flow of ideas.

"It isn't the form of government that makes life in this country what it is. It's the fact that technology has been exploited by knowledgeable people," declares Donald Alstadt. "There is a far better correlation between the standard of living and the overall well-being of a society with the degree to which the fruits of technology have been developed and distributed to that society than one will find if one correlates social well-being with the form of government, or the nature of the political or economic system involved."[2]

Britain, once a rich and powerful nation, now ranks at the bottom of the list of industrialized nations in productivity and per capita income. Yet, notes Lester Hogan of Fairchild Camera and Instrument Corporation, "she has continued to make her fair share of contributions to science and technology. England's relatively poor showing is associated with the political and economic system . . . which somehow has stifled her ability to make use of the products of science and technology for the welfare of her own citizens."[3]

A faltering pace of scientific and technological innovation suggests that the U.S. system isn't working as well as it once did. There is a growing body of evidence that tells us that laws and regulations have "caused a deterioration in the entrepreneurial environment," says Thomas Vanderslice, General Electric group vice-president. "They have resulted in a lack of incentives, both to innovators and to venture capital investors."

It is not the uncertainties of the future that bother businessmen so much as the attempts that are being made to eliminate uncertainty. The market system is built to take direction from failure. Some products fail. Some companies fail. Ours is not just a profit system, it is a profit-and-loss system. "Our capitalistic system depends on pains in some sectors so that adjustments are made," says David Potter of General Motors. "If we want a bland world, we would have to regulate everything."

Robert Colodny at the University of Pittsburgh raises a question about the motives for businessmen's cries about government's negative impact on technology. "Is management concerned about encroachment on management by government or about the loss of technological leadership by the U.S.?" Dr. Colodny concludes that "both are happening concurrently."

A No-Longer Free Market System

There are two basic ways of making choices and allocating resources: the free market economy or a government-directed economy. Others look to us as the last bastion of the free market system and lament that we are relinquishing it to a mixed economy in which government is increasingly interfering with or taking over the functions of business and individuals. Ironically, our system brought fantastic prosperity to the point where virtually anyone feels secure enough to express any grievance through the political system. And this, in turn, is distorting the economic system. In addition, some free marketers have abused the system and this has hastened governmental intrusion.

We may call ours a free market system, but it is not free. Robert Heilbroner, in *Business Civilization in Decline*, points out that there have been three major waves of government intervention in the U.S. The first was in colonial Ameria when government provided a direct stimulus for economic expansion. Then, after the Civil War, came the proliferation of agencies to regulate markets. And, beginning with the New Deal, government power was used to direct the economy to achieve goals such as full employment, growth, and welfare.[4]

Our economy, contrary to what some may think, has never been truly free. On the other hand, it is now approaching the point where it has difficulty functioning effectively as a *market* economy. We have come a long way from rugged individualism to expecting government to decide and provide. In an adversary relationship between government and business, innovation is suffering.

The dynamics of our system have not faltered because of some master scheme to make it fail although there are a few people who might like it to do so. It has been due primarily to the lack of understanding of how a private enterprise, free market system works. Society has hamstrung business, and the business community has not had a broad enough view of its role in society.

Neither a government-controlled economy nor a totally hands-off relationship toward business made this country, and neither of them is the mechanism for the future. Those who advocate the former can point nowhere to an example that would suggest that government control can encourage innovation or manage it in our best interests. Those who clamor for a return to "the good old days of free enterprise" don't know their history and haven't tuned in to the society in which they are living.

A New Social Contract

A dynamic, wealth-producing system that will benefit the maximum number of people must be built on the responsible person or corporation that serves society's needs and, in turn, receives the support necessary to continue producing benefits.

A great deal has been said lately about corporate social responsibility. Businessmen or people who own a piece of a business generally don't think of themselves as being responsible for playing a broad social role. Likewise, we hear of business's social contract, but people going into business were not aware they were entering a contract. For most of the nation's history, we thought "the business of America is business." Now, we talk of business's filling a role in society beyond providing what the customer wants in exchange for the opportunity to make a profit or earn a living.

We are stating a new condition—a worthwhile one, but a new one. Now that participation in our companies has broadened in terms of owners, managers, employees, and customers, business cannot operate in a world of its own. (And the public is no longer justified in referring to the "business world.") We are making the first attempts at defining a society in which business is a multipurpose part. We run the danger, however, of destroying business's primary capability and putting together a society without a productive engine. We may overlook the fact that a strong business sector that responds to society's expectations is better for all of us than one that is coerced. We could smother the will and ability to provide for our own welfare.

Profit Is Not The Only Object

Businessmen are rising to the new expectations of our society, however. Some of our most successful corporations are those who have shown exceptional responsibility in product, employment practices, and community relations. William Wendel, president of Carborundum Company, puts a new twist on our old standard—"We feel that the business of business is America."

Edward Ney, speaking as head of an advertising agency, may surprise and even offend some businessmen when he says, "My feeling is that business's necessary concern for profits can no longer be allowed to overshadow the need for preventive maintenance of the system which allows profits to exist in the first place."

Peccei, the Italian industrialist, goes a step further: "the social responsibility of the modern productive enterprise has become so all-commanding that it certainly cannot be sacrificed to the profit motive—which at the same time must be clearly recognized. Therefore, the first demand on any undertaking is definitely its social usefulness, around which, then its profitability can be organized—not vice versa."[5] Ian Wilson, a futurist and internal consultant with General Electric, says, "As a microcosm of society, a corporation must reflect all of that society's shared values—social, moral, political and legal, as well as economic."[6]

Whether the corporation is organized first around profit or social usefulness, the realities are that both will have to be served. Exxon USA President Randall Meyer says, "the interests of business and interests of the public in all the aspects of a better quality of life are inextricably intertwined." He, like many other industrial leaders, recognizes that managers of business have done a poor job of explaining business realities to the public. When they do speak up, they tend to use cold economic terms that "reinforce the misperception that management is interested in short-term profits and nothing else."

Social, economic, and political trends of recent decades have brought industry and government closer together but in an adversary relationship. While there is a broadening interface between them, they too often push against each other rather than together toward creative solutions to social and economic problems.

Is A Partnership Possible?

"Does the dogma of old, which conceptually if not physically fostered separatism and compartmentalization of societal, political, and business interests, need review and revision?" asks George Dominguez of CIBA-GEIGY Corp.[7] It clearly does, he believes.

We have to balance "the competitive enterprise system from which innovation, entrepreneurship and new products and new processes come" with "a sensible series of restraints," says Herbert Hollomon.[8] Some industry managers look for improvement in the relationship between government and business. They even hope that the relationship could become more of a partnership than an equilibrium between adversaries. This does not mean, however, that business will be released from its social obligations. Management's concern with impacts from outside their business is not going to ease up, says Neil Drobny of Battelle. "The momentum is unstoppable because of society's raised expectations. It doesn't matter whether these expectations are realistic or not."

Two trends are occurring which could lead to a partnership. In government, there has been an increase in the number of people operating with a more professional management style. Although businessmen are unhappy with federal, state, and local bureaucracies, they admit that they have seen a rise in the number of well-intentioned, enlightened people in key spots. Someday, they hope, an avenue of manager-to-manager relations may cut across government-business lines. This would facilitate the attainment of society's objectives and lessen the danger of ignoring economic realities. Businessmen are also hopeful that more technically oriented people will find their way into administrative agencies and legislatures.

In business, authoritarian management is giving way to more participative forms which are sensitizing business to society's needs. A major influence on the definition of the corporation's societal role will be exerted by changes in ownership. The concept of a shareholder being the little old lady in tennis shoes, with stock in a company and concerned only about retirement income, is fading into the past. Increasingly, stock is held in blocks by pension funds, insurance companies, and other institutions. Today, says Max Sporer of Touche Ross, "it seems the power has swung from those who want profits to those who want something else." In the years ahead, these groups are apt to become educated in economic trade-offs, however. Rather than forcing management to stew in a pressure cooker, trying to deliver profit growth, a better environment, safer products, improved working conditions, and more fringe benefits, the owners themselves may have to face up to these trade-offs. "It'll be management's job to lay out the alternatives," says Sporer. If this is the case, business managers may trade their adversarial role for that of mediator.

The Reward For Risk

Profit may be seen in a new context by managers, owners, and society in general, but it will continue to play a part in the allocation of our resources and investments. Profit is a driving force that invites men to take risks and to adapt to change. For only a moment in the age of man did it flow to those who happened to be in the right place at the right time and who could ignore the conflict around them. As we look to the decades ahead, it seems far more likely that it will be won by those who put themselves in the right place by responding to society's needs.

If funds are going to be applied to develop innovative solutions to our problems and to explore new opportunities, there has to be a reasonable expectation of profit. Regulations that raise the costs or limit the potential return diminish the capability to use our know-how. The more funding we generate through taxes, the less there is in the wealth-creating sector to risk on new ideas.

The U.S. has a severe problem in the taxation of capital, says Dr. Noyce of Intel. "Because it is so heavily taxed, it isn't available for new startup companies, and there is much less interest by the individual to invest for capital gains." A capital shortage will be one of the most important issues of our time, say some industry leaders.

The New York Stock Exchange released a study in 1974, showing the savings potential for the U.S. was $4 trillion through 1985 compared with potential capital needs totaling $4.7 trillion. At that time there was considerable concern among businessmen over their chances of getting funds over the long term for expansion and modernization. The concern subsided during the recession of the mid-seventies. But it will rise again when business activity reaches the pressure point at which companies think of big spending programs. Some companies—particularly the smaller ones—will be shut out of the market for funds.

Despite the lack of investor interest and the bias of tax policies against savings and investment, the business sector feels challenged to raise massive amounts of capital to prove it can do its job. Some companies will have to venture boldly into delivering "big projects" that will prove industry's capability to bring technology to bear on major problems such as energy, natural resources, health care, or food, says Ed David. He visualizes $100 billion-plus projects financed by industry—not government. Management will have to find ways to get together to raise the capital, deliver the answers, and find a profit in it, he says.

When businessmen complain about the lack of adequate capital, what they are really referring to is the out-of-balance risk-reward ratio. Capital isn't coming forth because the odds against getting a reasonable return are too high. A large part of the problem is due to uncertainty, says William Ylvisaker. "A lot of people are unsure of what the restrictions are going to be—taxes, credits, regulations, etc. Because so many regulatory agencies impact on business, there is tremendous insecurity." In 1975, in hearings before the Subcommittee on Economic Growth of Congress's Joint Economic Committee, Jerome Wiesner, president of MIT, pointed out that "government influences, and sometimes even dominates, the market decisions so that it is not possible for a businessman to judge the outcome of an investment he is contemplating."[9]

"No Policy" Has Become Policy

What has been the cumulative effect of national goals and regulations on our technological /creative capability? The Commerce Department, in a study on the need for a technology policy, has stated: "A coherent national technology policy needs to be developed in order to maximize the U.S. capacity to develop and utilize technology to achieve national purposes. Every proposed national policy, whether or not obviously technology related, should be evaluated for its potential impact on technology," the study concluded.[10]

As we make decisions in regard to foreign trade, employment, health care, and hundreds of other aspects of our national life, we sometimes do so without considering the harm done to our creative capability and spirit. Businessmen aren't asking for a national technology policy but some feel the need for policies to serve particular industry sectors. "There are many cases that will never fall within the purview of specific institutions or companies. These are national goals," says Xerox's Jacob Goldman. "In these areas—energy, to name one—government has to and will unquestionably establish broad policy. But if you narrow it down too fast, focus on what somebody thinks is the most promising avenue, and neglect all the other options or even the basics and fundamentals that may lead to new options, that's where too much government involvement will inhibit solutions."

Government should not try to determine the course of science and technology, says Dr. Goldman. It can help best "in providing a certain amount of motivation to provide an environment in which the problems that government is interested in solving can get solved." Government could employ such devices as tax credits, other tax incentives, liberalization of the patent system, and special help to industries whose assets are locked into obsolescent technologies.

"In peacetime, in a democracy, government's role is not leadership," says Dr. Fawcett of Battelle Memorial Institute. This is the responsibility of the private sector. And that has to originate in people's personal ethics, he says. We have a 55 miles per hour speed limit, for example. It was legislated—not for safety—but for energy conservation. We can't police it thoroughly, however, without putting so many police cars on the road that we would waste more gasoline. The objective of conserving energy, therefore, has to be a personal ethic, Fawcett maintains.

Total reliance on the market system exposes two problems, however. It may tend to make decisions on a short- rather than a long-term basis, and it could neglect social values. Schumacher points out that the market "represents only the surface of society and its significance relates to the momentary situation as it exists there and then." It is "institutionalisation of individualism and non-responsibility."[11]

This may reflect a pessimistic view of man but a realistic one until such time as the buyer and seller each accepts responsibility for more than himself. At the same time, however, the governmental process embodies that same lack of responsibility; it, too, is the summation of selfish demands and shortsighted responses.

Government can only support what it understands. If innovation does not come from the private sector and the individual, it cannot be forced out by laws and regulations. The authors of a series of white papers on technology, sponsored by Gould Inc., state: "Innovation cannot be dictated by the state. Rather, it rises from an environment that respects the insights of everyone—and judges ideas on their merit, their usefulness, and their acceptance by the marketplace."

Perhaps business and government, working in a creative partnership, can capture the best insights and promote their acceptance. Personal and societal goals could be blended together. Business could use its technological and marketing skills, for example, to create fashionable clothing, homes, and transportation systems that also conserve energy. It could direct its powerful persuasive efforts to elevating conservation as a personal ethic—possibly making it contribute more to sex appeal than toothpaste. Quick, one-shot products or promotion campaigns won't bring about such change. Bringing personal and societal goals together will take years of fashioning a new frame of mind. But business has the advantage of being future-oriented. The continuity and long-term stability of the corporation could keep us on track toward goals that might fall victim to changes in government.

Government's Proper Role

"Government's proper role in society is to solve problems which individuals and private firms operating in a free, competitive marketplace do not have the incentive or means to solve on their own," said John Byington at the 1976 Forum on Business, Government, and the Public Interest. "Government must act as a catalyst, a stimulant, a motivator—working in a creative partnership with the private sector to solve the environmental, health, safety, and other problems of our complex society," the Consumer Product Safety Commission chairman declared.

In a 1977 speech, Elmer Staats cited some examples of government's participation in economic and technological development: establishment of the patent system, policies to support agriculture, grants of land and rights-of-way to aid development of railroads, investment in highway systems, support of R&D which led to the electronics revolution and world leadership in aviation.

A blue-ribbon panel on invention and innovation, reporting to the Commerce Department in 1967, concluded that "it is incumbent upon government, both local and national, to provide the essential framework for social innovation." It said government should encourage the use of private resources for innovation whenever possible. The report described government's role as:

"a. Defining the social problems and the priorities for their solutions.
"b. Intensifying the planning for such solutions.
"c. Encouraging private enterprise to seek profit-making opportunities in the development of such solutions.
"d. Developing regulatory and other mechanisms, such as government purchasing policies, to compel or encourage industries to modify productive processes and products in such ways that they will contribute to the betterment of the social sector.
"e. Carrying on the necessary technological developments when it is clear that private resources cannot be depended upon to undertake them satisfactorily."[12]

While Dr. Betsy Ancker-Johnson served as assistant secretary of Commerce for science and technology, she was a strong advocate of government promotion of industrial technology in the manner it has promoted agriculture. She favors funding research on generic problems that would help large segments of industry. Through government grants and by permitting companies to join in consortia, work could be done on such common problems as corrosion, combustion, and materials substitutes.

From about 1940 to 1970, "federal funds were the prime energizing force in the nation's laboratories," says GE's Thomas Vanderslice. Before 1940, only 20 percent of R&D support came from federal funds. From the early fifties to the mid-sixties, government provided 60 percent of the dollars. "This fortuitous set of circumstances began to fall apart in the sixties and early seventies," says Vanderslice. "The clean sweep by the U.S. of the 1976 Nobel prizes in the sciences and the two out of four in 1977 is tribute to our past—not necessarily a sign of the future."

The Trouble With Government R&D

The consensus in industry is that basic research, in particular, should be supported primarily by government funding. Business managers abdicate this area of R&D, insisting that their concern should be farther down the innovation line closer to product development. President Carter's science adviser Frank Press says government has a special role in basic research because its distant payoffs are for the good of the country. "The basic research will end up eventually being used in important technological developments, but it's hard to predict that in advance. It's hard to patent the results of basic research, so industry is reluctant to make major investments in it. Industry people are very supportive of the money the government puts into universities and basic research centers because they know that, eventually, they will benefit from the new knowledge generated."

Betsy Ancker-Johnson stresses that government research need not be in the form of gigantic programs; well-placed dollars could have big payoffs in savings for industry and benefit the entire economy. Unfortunately, however, legislators and government administrators have a tendency to think on too-grand a scale. Government agencies often exhibit the belief that bigger is better. One innovator complains, "you can propose a modest funding for an effective program but they prefer to get into something bigger."

Government involvement in research can be capricious. Funding is generally on a short-term basis and continually subject to election-minded officials. Technological solutions to many of our problems take a great deal of time. One reason the public is often disappointed in R&D is because "the development cycle runs longer than the political term of office so you end up with failure," says Egils Milbergs. Political pressures can lead to "crash funding" of programs even when they are not worthwhile, he adds.

Breakdown in the continuity of a worthwhile program, on the other hand, can be costly. Lack of it stretches out the costs and reduces the present worth of future gain, says GE's James Young. It's also needed if the innovators working on a program are to maintain their enthusiasm.

In a paper published by the Center for the Study of American Business, Roland N. McKean raises the question of whether basic research can remain basic research in the hands of government. "It is difficult to rig costs and rewards within government so that relevant officials can capture gains from pushing desired basic research." Since the public isn't clamoring for it and elected officials don't gain much from supporting it, department officials "are constantly tempted to shift R&D funds toward applied or engineering developments."[13]

While business has traditionally admitted its primary interest in R&D is the development of products and processes with payoffs that are not too far into the future, it has assumed that government can afford to engage in longer-term research. Yet, both the public and the politicians are looking for measurable results and they aren't willing to wait a long time for them. In their concern for relevance, they may limit the options by failing to develop knowledge across a broad base.

Leaders in science and industry advise that any government ventures into R&D be relatively small since programs may be overfunded and overextended when government officials and private interests find them an easy source of votes and money. They advise, too, that government stay out of the latter stages of development work. It cannot provide the linkup between discovery and the marketplace as well as business does. Government should only advance R&D up to the point of commercial development and not decide which technology is to be developed, said professor Robert Gilpin of Princeton University at the Joint Economic Committee hearings on technology and economic growth in 1975. He pointed out that when market considerations become the key consideration in a development, the coupling of the new technology and the market is "best done by industry." He suggested that government "create the incentives and disincentives which will encourage industries to be more innovative in their use of their R&D resources."[14]

Pay Attention To Industry's Innovation Capacity

We should be looking at each vital sector of the economy and determining how to improve its innovative capacity. While we can, through government, indicate where we want our innovative capabilities applied, it is impossible to guess from what quarters the answers might come. A neat, coordinated, or controlled approach to R&D might look efficient but it is unlikely to be effective in the long run. Government has proven effective in coping with specific projects such as defining a moon-landing mission and bringing the talents to bear on it. But our base of scientific and technological know-how is a complicated weave of major and minor innovations coming from individuals, corporations, and government—both here and abroad.

Business would like government to be more helpful in the transfer of information. William Norris of Control Data Corporation estimates the federal government spends at least $1 billion a year to disseminate results of federally funded R&D. Yet, he claims, it is extremely difficult, if not impossible, for industry to use the information available because of "inadequate records, ineffective coordination of federal programs, lack of attention to the needs of industry."

Norris suggests that technology transfer be encouraged by stimulating government laboratories and universities to make their information available, encouraging agencies to make their data bases available at minimal cost, and creating incentives to encourage private companies to sell or lease their technology. Frank Press agrees that the transfer could be improved but feels that it's better than in most other countries. Cooperation between industry and universities is better here, for example, says Dr. Press. He is encouraged by the "novel experiments of joint funding by government agencies of industry-university teams."

The Market System Needs Help

Government's role in an innovative society involves discussion of more than the issue of its participation in research and development. All its activities and policies should be examined to determine their impact on the innovative capability of business. The entire innovation process, not just R&D, needs careful tending to ensure that the necessary incentives are working to keep innovation alive and directed toward society's objectives.

Looking to government simply as the funder of R&D is an "undesirable direction," says Milbergs. "The private sector gives up. It says, 'Well, you can't raise the capital, the risks are too high, the regulatory uncertainty is too great.' It says it can't deal with society's problems—'let the government take care of it all.' What we really need is a bunch of experts who understand how the marketplace runs."

The market system has to be the vehicle for innovation, but it needs some government involvement. One such involvement is suggested by Milbergs—organizing the marketplace, which has become "more of a social marketplace." He explains, "General Electric doesn't sell nuclear power plants to a utility; it is really selling them to the community." Mass transit is another area in which the community is the market. But the market structure is chaotic since each transit district has its own goals, problems, specifications for equipment, and procurement procedures. This is the sort of market in which federal guidance might speed up the application of technology to provide less expensive and better solutions. Instead of putting its money into R&D, says Milbergs, government could subsidize the buyers of the system in an effort to rationalize procurement specifications and standardize products.

Some people believe that federal procurement policies could be used to influence technological innovation. Although it was *the* market for defense and space goods, the federal government represents a small share of the market for most civilian goods. And, when purchase orders are placed with the supplier offering the lowest initial cost, the best technology is not always encouraged.

Government should find that the best way to implement its policies and promote innovative responses to societal problems is through the market mechanism. Milbergs makes an important distinction: "Private industry is market responsive; it hasn't been policy responsive," Policy, for example, might be saying "conserve energy" while people—or the market—are saying "consume more energy." If government can find ways of translating its policies or objectives into incentives in the marketplace, industry will respond with all its innovative forces.

"It is not the goals but the specific means to be used that most often provoke conflict between business and government," observes Irving Shapiro. The du Pont chairman urges businessmen to make their expertise show more convincingly "by offering our best, most objective counsel on technical, financial, and managerial facts.

"We must go to Washington early-on with an unemotional, documented case of facts supporting our position, leaving behind any free enterprise lectures or ideological sermons. We'll make headway. When I meet with members of the Congress or the Executive branch on pending issues, I get a good hearing. I also find that the members, on many occasions, have not been completely aware of all the facts involved and that they are anxious to learn all the facts before decisions are made. When legislative and executive actions are based on the facts, they reflect more of the engineering, managerial, and financial realities of the real world. That condition should be our goal."

14 Dialogue and Control

"Technology is the expression of the society; it is an expression of the values and the abilities of the people that generate it. It is indeed a most revealing indicator of our society. And the fact is that technology in turn shapes the values of a society and of its people."

George Bugliarello
President, Polytechnic
Institute of New York

"We see in the U.S. today a severe mismatch between the high potential of technological advance and the slow pace of the country's social-political progress," says Simon Ramo, vice chairman of TRW Corporation and well-known adviser to government on technological matters. The reason we are not making full use of science and technology is because we are weak in organization and cooperation, Dr. Ramo says.

"What is known in the world and what is possible in term of technology is definitely not being applied as effectively and as widely as it should," says Donald Alstadt. The gap exists, he asserts, because society does not understand what is technologically possible now.

Ramo and Alstadt are scientists and business managers troubled by awareness of the difference between what is and what could be. They see the need for a bridge of education across the gap between a society with problems and technology with potential solutions. If such a bridge is to be built, it will have to be a two-way structure to serve effectively.

From one end of the bridge will come scientists, technologists, and managers whose perceptions of technology differ from the public's. They will have to recognize people's apprehensions about the nature and pace of change and the risk technology poses to jobs, the environment, and life itself. In order to direct technology towards society's needs, "we have to have a very strong scientific community that is sensitized to the fact that there are problems and that listens to them," says Xerox's Jacob Goldman. The public's disillusionment with technology calls for action on the part of the technological community, says Clark Abt, in *The Social Audit for Management*. "First, technologists need to develop rapport with social scientists and learn from them what the major social needs are . . . Second, technologists must not give up being technologists because technology cannot be applied to social probems if technology is not continually being developed. Third, technologists need to become sensitive to the impacts of major technological changes on the quality of life and to determine if such impacts are consistent with a publicly acceptable social role for technology."[1]

Better management of its creative people, better management of its dollars for innovation, better solutions for its customers' problems are needed if industry is to fulfill its traditional role. But even all this is not sufficient. "The new realities will require that business be managed by leaders who are actively in tune with the larger goals of society," says Irving Shapiro. "If we want to retain our basic mission of being primarily responsible for our society's goods and services, then we must work on that mission with positive evidence that we are operating with the society's new expectations as firmly in mind as our corporate objectives. We must demonstrate by deeds that we are not a self-serving force that regards society only as a marketplace."

Garbled Messages

Members of the business community are concerned because society does not appreciate them or understand the economic system. In their limited efforts to "sell" the public on the private enterprise system, they say, "look at the goods and services we bring you." The public shouts back, "You also bring us inferior products, health hazards, pollution . . . . " A handful of business and scientific leaders have begun to speak up in the last two or three years about the slowdown in our pace of technological innovation, but the public has paid little attention to what it thinks is untrue, irrelevant, or—possibly—good news.

The messages are not making it across the gap. The business /scientific community can't even get together to construct its message. In the area of economic ideology, there are some statesmen who see industry in a broad societal role, but there are far more who take a limited view of why they are in business. The leaders who speak up about the slowdown in innovation, particularly in fundamental discoveries, are outnumbered by those who are pleased with their efforts in product design—what Jacob Goldman calls "improving the old" rather than the big breakthroughs. Managers, engineers, and scientists don't like to hear that they are associated with a decline; after all, they are doing their best and they have done well. Basic research is for the universities and government anyway, so if it is declining that's their problem, these people think.

The scientific process has always employed an adversary process in order to get at the truth, Arthur Kantrowitz explains. But the public isn't aware of that; it simply observes that there is disagreement. Kantrowitz warns his colleagues that there is no way of communicating between a divided scientific community and the public. "To tax the media with all of the confusion that we have today is too much. If we straightened out our own house, if the scientific community had a way of credibly communicating with the public and offering that kind of opportunity to the media to understand the issues, then we could start blaming them for the confusion . . ."[2]

The communications media find an adversary relationship good fodder for stories. Scientist versus scientist or scientist versus the public may make great "copy," but it does not always lead to understanding. Even the tape-recorder-equipped, newspaper investigator or the TV reporter leading a phalanx of technicians is susceptible to the ease with which he can generate excitement by taking a negative approach to technology or business. Ironically, the news media people are highly protective of the right to use their technology in absolute freedom.

Society does not stand to profit from an adversary relationship. To the extent that we have seen some progress in tearing down these walls, "it has been mostly on the side of the scientists," asserts Mr. Shapiro. "They have climbed down from their tower to concern themselves with the social and economic and ethical consequences of their work. The reverse is not so true; the nonscientists have not met the scientists halfway. This is particularly true in government where many of the people in positions of high responsibility are lawyers. As one member of our profession has put it, the law is 'one of the last strongholds of the scientifically illiterate.' "

Telling The Story Of Technology

Businessmen, scientists, and technologists have to tell their story if they want to restore the climate in which innovation once thrived. "We have to speak out more about what technology has done—and undone," says William Ylvisaker. His own firm, Gould Inc., has produced a landmark series of "white papers" which relate the history and importance of technology.

Even a relatively small company can do a lot to foster a better understanding of technology. Closely held Lord Corporation cosponsored "The National Symposium on Technology and Society" in late 1977. Together with WQLN, a public TV station, it lined up top speakers from industry and science. An outgrowth of the forum was the establishment of a task force to coordinate and fund activities to promote a better understanding of technology and its role in society.

The public may be more receptive to the message than we expect. Although the opinion polls would lead us to believe that there is a crisis of confidence in science, the facts show that "ambivalence, not rejection, best describes public attitudes," says Clyde Z. Nunn, senior project director, Newspaper Advertising Bureau Inc. Readership studies indicate that newspaper readers have a high interest in energy problems, social problems, environment, nutrition, and health, he says. But "fewer science-related items appear in newspapers than the interest ratings indicate should be there, and only 11 percent of the daily newspapers in the nation have science editors."[3]

The scientist, technologist, or businessman must guard against talking in terms that relate to his efforts but fail to convey the meaning of those efforts to others. Public officials are receptive to learning about technological developments, problems, and possibilities, say some business managers. While many officials are locked into certain positions, businessmen have found that there is a significant group in the middle who are genuinely looking for facts. "We are in a technological society, but our representatives are not technologists," says Thomas Miller of Union Carbide. "We have to interface and provide factual information." In regulatory matters, others have found, scientists and managers are welcomed when they bring information on costs and consequences of a proposed rule, especially if they don't give the impression that they are making a judgment on the rule.

Industry and the scientific community are in the best position to present information on the potential consequences of certain technological steps. The American Society of Mechanical Engineers, in its fall 1977 newsletter, said dialogues between engineers and lay people could help combat strong antitechnology feelings. These dialogues, however, "must not take the form of the engineer defending unlimited technology, but rather the engineer explaining to the nonengineer the concept of unintended effects and helping the lay person to weigh, for a variety of products, open and hidden disadvantages as well as advantages."

Monsanto Company has not only generated a considerable amount of communications about the risks and benefits of technology but has made available $500,000 to support the advanced training of toxicologists, whose work is to study risk. The company has also entered into a joint research project with Harvard University that dramatically reflects its involvement with social goals. Monsanto is putting up $23 million for a 12-year program to study the vascularization of cells—advanced research that could lead to a better understanding of the causes of cancer. The company will furnish the R&D facilities and people to back up top university researchers. It calls its relationship a "window on biology"—an opportunity to work with academia where most basic research on human health takes place.

This suggests one further step in bridging the antitechnology gap. Industry and the scientific community should help the public visualize where they are headed—what future capabilities they are striving for. They must relate this to the quality of life and then let the people know that they—not a technological elite—will determine the direction of technological progress.

The Public's Responsibility To Learn

Not all the bridge crossing should be done by scientists, engineers, and business managers. The public and its government officials, have a responsibility to learn more about the potential and the limitations of technology. They need to understand the nature of scientific disciplines and the sense of achievement that drives some people to seek the truth and to create.

We live in a technological society. Although we cannot each be an ecologist-computer designer-chemist-physicist, we do have to know enough about technology to know what it is not. We must know what questions to ask the specialists. We do not design our own cars—few of us can even repair them—but we do drive them. And we do not fear them. This same courage and trust has to extend across the technological spectrum. We must ask the technologist "What are you doing? What will it do for me?" And if we don't like the answers we hear, we say "stop." The more enlightened we are, the less apt we will be to call a halt to worthwhile developments out of ignorance, selfishness, or shortsightedness.

Environmental protection is a good example of the challenge to understand the problem in order to resolve conflict and find good answers. The key to government's success lies in "understanding the nature of the problem and the consequences of proposed solutions," says J. William Haun, vice-president, General Mills Inc. Much of the environmental discourse "is couched in scientific and technical language," he notes. "Government officials, even diplomats, had better begin to understand the difference between criteria and standards, between a toxic substance and a hazardous material, and how the atmosphere circulates over the continents, because they are essential items of knowledge."

In simpler times, a person could master most of the technologies around him. "Today, in contrast, the technology that sustains our existence is known only to a tiny corps of specialists," says Gerard Piel in *Science in the Cause of Man*. Even these people know only one narrow branch of science. "The public issues that preoccupy the political life of America today are heavily conditioned by purely scientific considerations involving realms of information and ideas that are unfamiliar and unknown to the vast majority of the American People."[4]

One might wonder how we can democratize the guidance of technology if people do not have adequate knowledge and can't acquire it because they don't speak the language. There is another side to this coin, says June Goodfield, adjunct professor at Rockefeller University. That is "the scientist's understanding of the public. Science and society must be closer to one another." She urges scientists to find ways of expressing the humanity that has always been present in science as they respond to the new imperatives in society. Society, in turn, must find new ways to understand them and to help the scientists express this humanity.[5] The scientist or engineer who brings us an innovation is human, sensitive, and imaginative. But we see him, all too often, only in terms of his product. Together, we must create a relationship in which he will share with us his problems and hopes. If the public can learn something of the language of science and the workings of the technologies around them, so much the better. But technology is a human endeavor and it is only in human terms that we will clear the language barrier.

A Challenge To The Educational Sector

The rise of public involvement in technological issues contrasts sharply with its preparation to do so, agrees James Young of General Electric. One area on which he would focus attention is university education. "It seems clear that technology must become a fourth stem of the liberal arts education to insure better preparation" for the trade-offs society must make. "Equally," he says, "the humanities stem of engineering education must consolidate itself in ways that can serve the new social and political responsibilities of the technologist."

Simon Ramo has frequently called attention to the engineering profession as one of the most influential factors in determining the effectiveness of science and technology as servants of the public interest. Up to now, he says, engineering has generally been thought of as being concerned with designing and building machines and systems. But this definition has become inadequate.

"The proper use of science—its timely and wise application to help man with his problems, enhance his opportunities, provide him with acceptable options, and satisfy his social and economic requirements—now is seen to constitute an endeavor of vast proportions," says Ramo. "It is this broad endeavor, the overall matching of scientific and technological advance to social needs and progress that must constitute 'engineering.' "

Engineering schools attempt to inject some liberal arts into their curriculum. Liberal arts colleges experiment with strengthening the science content of their programs. Yet, the two "cultures" go their different ways, concerned with quite different interests and disciplines.

Changes in curricula could help humanize our uses of technology, but the educational sector cannot do the job alone. June Goodfield would dispel any notion that the humanities and the study of the great thinkers will automatically give us moral virtue. "I think it unfair and unwise to regard the humanities in this therapeutic light. They are good in themselves and should not be regarded as remedies for our own failings. We must not pretend that words and university courses are a substitute for human hearts and human action."[6]

Most of our educational system is geared to preparing people for jobs, giving them little of the analytical skills or life-awareness of the liberal arts or the disciplines of science and engineering. "We haven't succeeded in giving people the ability to live better or to find satisfaction in life," says behavioral scientist Fred Herzberg. "The university is no longer a collegium of scholars; it's a ticket-punching system in which people qualify themselves for positions."

In the years ahead, the entire educational system—not just universities—will be challenged to prepare people for managing a technological society and for living full lives within it. It will have to foster an appreciation for the innovation process and be more visionary of alternative futures. True education opens the doors to change, without which societies become ritualistic and static—and die.

Democratizing The Control Of Technology

The public not only wants to know what science and technology are doing to affect the quality of life, it has a general fear when it does not know what research and development people are working at. They want to know what harm can come from the end results or from the experimentation itself.

A bit of fear of the mad scientist or a Frankenstein monster colors people's attitude toward the laboratory. At a laboratory where government work is underway, access is stringently controlled in the name of national defense. In corporate laboratories, strict security protects proprietary information.

In *Premeditated Man,* Richard Restak quotes British biochemist and brain scientist Steven Rose on the matter of controls in the field of biomedicine. "What controls exist now are in the hands of the powerful members of society. This leaves us with two problems: how to democratize knowledge and how to democratize these controls." Rose puts most of the burden of democratizing knowledge on the scientist, who must not engage in secret research. Restak says that Rose and other scientists are advocating that the scientific community "let the public in on the decision from the beginning." This would be a reversal of the present pattern of making decisions without public input, encountering disaster, inviting public alarm, and then being set upon by "watchdog" institutions.[7]

The public cannot control what it cannot see. And, yet, how can it be shown everything? What purpose would have been served if the Manhattan project—building the first atomic bomb—had been proposed first to the public? Why would a company put money into R&D if it is not allowed to gain an advantage over its competitors?

These are questions we will have to resolve because the ascendance of public involvement runs counter to the secrecy that is still supported by political and economical realities.

The Prime Influence: Society's Values

The innovator who is concerned with the social consequences of his work doesn't expect unrestricted liberty. He lives within and, in fact, serves the values of society. The best control on science and technology is his self-control. The nature and pervasiveness of self-control may be a function of the society in which the innovator works, however. Horrible medical abuses were perpetrated in Nazi Germany because they were tolerated by people who knew of them and encouraged by those in power.

Much of man's scientific and technological progress has been initiated by the demand for the tools of war. From the time man first used a bone or branch for a club, his tools extended the potential for harming his fellow man. The spear, bow and arrow, musket, airplane, rocket—all extended his dangerous reach. In fact, his reach has become overreach for some purposes. The weapons of post-World War II became so powerful that the U.S. struggled to resurrect old aircraft and design new ones that were slow and maneuverable enough to match the pace of small-scale combat in Vietnam.

The problem of war is a social problem, not a technological one. We once thought that weapons had become so potent that they had ended war itself, but we now know better than that. And, while we hope we can control arms and limit the availability of the ingredients for horrible weapons, we can see that the materials and know-how for constructing them are becoming accessible to more and more nations and to terrorists. We cannot hide our technology; we can only work to eliminate the causes of war.

A primary control of the scientist has been exerted by his fellow scientists. By voluntarily submitting to peer pressure and helping fashion regulations that influence objectives and govern procedures employed in his work, the scientist has done most of the controlling himself.

Researchers in the field of recombinant DNA are opening the doors to knowledge that may eliminate certain diseases and enable us to improve on man's body. This knowledge may also be used to create monsters and unleash diseases we have never known. The researchers have pointed out the potential consequences and raised numerous ethical questions. We should not overlook the fact that "the entire recombinant DNA fracas was born of a unique act of social responsibility on the part of the scientists involved, when they voluntarily halted their work and tried to evaluate its potential risk to public health," says Robert M. May of the Princeton University biology department.[8]

The scientist or technologist can also be influenced through the pocketbook. An innovator with no funds cannot do much, especially in this age where progress often depends on big investments. To a large degree, we can direct his work through our roles of taxpayer, stockholder, manager, and customer.

Despite peer pressure, standards of professionalism, and control of the pursestrings, it is possible that antisocial developments can still occur—deliberately or accidentally. We might, then, be inclined to seek more and more rigid governmental controls. But we cannot eliminate all risk either in the work of the innovator or the use people might make of the ultimate end-products. No control is a sure control. Absolute controls would be a severe, if not fatal, depressant on inquiry and innovation. They run the risk, too, of turning science and technology over to autocrats.

This is the dilemma we face. Our history suggests that we should opt for freedom and bear the risk of the minority who might abuse it. We cannot base a science and technology policy on the premise that all scientists and technologists are guilty until proven innocent.

"Methods for imagining, communicating and comparing comprehensive alternative futures are not fully developed," admits professor Carl Madden. "But, the problem for leadership is to create, through participation and active contribution, a shared perception of the new culture which is needed if mankind is to avoid the constraints which threaten the exhaustion of creative effort, or social elitism with its threat of authoritarianism, on the one hand, or passionate revolution with its threat of authoritarianism, on the other."[9]

Technology can only serve humanistic ends in a society of humane relationships—both personal and institutional. June Goodfield points out that ". . . if we insist that the scientific profession and the medical profession have a care and a human concern which we ourselves as members of society are not prepared to have or to act on, we shall be raging hypocrites."[10]

Industrialist, scientist, engineer, news analyst, educator—each shares the responsibility to promote an understanding of technology. Difficult as that task may be, the transmission of the public's message in the reverse direction poses an even greater challenge. That will necessitate establishing a cohesive set of values and priorities toward which we want our technologies applied. Until both sets of communications are flowing clearly and continuously, there can be little protection against irresponsible acts of technologists or the users of technology. There is no simple, quick solution for controlling either the positive or negative effects of technology.

15 If Man Is To Be Man

"Man, in the unsearchable darkness, knoweth one thing
that as he is, so was he made: and if the Essence
and characteristic faculty of humanity
is our conscient Reason and our desire of knowledge
that was Nature's Purpose in the making of man."

Robert Bridges
The Testament of Beauty

Man took the fire in a dark world and made himself a home. His own intelligence advanced, casting light on his world and himself. Nothing could resist his consuming drive for knowledge and freedom. Few things escaped being refashioned in the glow of his creativity.

Material wealth and comforts, won through the know-how we have acquired, can be readily appreciated, even if we do tend to take them for granted and occasionally need to have the relationship called to our attention. If someone points out that, through ignorance, fear, or indifference, we are jeopardizing their availability, we might scurry to take some remedial action. Yet the slowdown in our pace of technological innovation is not regarded with universal regret because some people think we have had enough economic and technological growth or more change than we can manage.

The attainment of wealth and comfort is a poor basis on which to try to sell technology. Trying to rally allegiance to technology on the basis of its contribution to man's survival is not a much stronger argument since he is battling not only the natural environment but the one he created with his technology. He still struggles with his fellow man, and the conflict grows more frightening because of his extended powers.

There is a better explanation for the persistent burning of the technological spirit when man's own indifference and hostility threaten to extinguish it. Although man has dramatically altered his position in the world and increased his comforts, he is not satisfied. His technology has nearly wiped out starvation in many parts of the world, yet he hungers for something. Technology has added layer upon layer of wants as it has satisfied others. The progression is endless, but the pursuit goes on.

It has been ambition, not laziness, that has driven man to make improvements on his world. First, he fought to survive; then to find identity and self-fulfillment. He searches not simply for satisfaction for man the animal but for significance.

The information imparted through man's genes constitutes a decreasing portion of his total knowledge. As man ascended, the development of brain permitted the storage of immeasurable amounts of information developed individually and collectively. He has set up knowledge storage facilities outside himself in books and computer data banks. This information extends well beyond his survival needs.

It could be argued—particularly in the more fortunate nations—that we have enough science and technology to satisfy our material wants, writes Gerard Piel. "If the advance of technology supplied the sole motive for work in science, then the book might be closed without regret. But, of course, this is not the case, and the work of science is not yet done. To bring the work to a stop, man must be made to stop asking questions."[1]

The flames illuminate more and more of the world, and man insists on peering still deeper into the darkness in which he awoke. As Daniel Bell expresses it, "boundless ambition has ruled the quest for knowledge. At first man sought to conquer the natural order; and in this he has almost succeeded. In the last hundred years he has sought to substitute a technical order for the natural order; and in this he is well on his way." But now, when knowledge is growing exponentially, says Bell, "there is a question of whether man will or will not want to proceed."[2]

"The Irresistible Need To Explore"

Technology, measured in terms of products and physical comfort can lead to an improved quality of life. But this is only a two-dimensional view of its importance. The inquiry and adaptation it embodies, particularly when led by scientific investigation, is the important third dimension. Science has utility value but it has been, from its birth, a natural philosophy—a method of inquiry. It is the torch man carries in the search for his identity and his relationship to the world—who he is, his origin, his role in the universe. It has increasingly become the foundation for technology, and technology, in turn, assists the advance of science.

Is all science and technology aimed at determining the future? Are we proceeding with undue haste toward some future where we have no business being? Man might point out that he has had to make it his business to be on this planet at all. He has had to explore the past as well as the future to earn what limited security he has here.

We assume that everything pertaining to space exploration, for example, deals with the future, says William Birenbaum, president of Antioch University. But that is not the case. The farther we reach out in space, he says, the more we reach to the past. The thrust of astronomy today is to learn about the beginning. Light coming from the outer edges of space comes from the distant past—not the future.

If we fail to see science and the application of it as part of our evolution, we miss the heart of it and deny our own humanity. Jacob Bronowski wrote: "We are nature's unique experiment to make the rational intelligence prove itself sounder than the reflex. Knowledge is our destiny.[3]

"The body of technical science burdens and threatens us because we are trying to employ the body without the spirit; we are trying to buy the corpse of science," said Bronowski.[4] "What science has to teach us," he concluded, "is not its techniques but its spirit: the irresistible need to explore."[5]

Two Perceptions Of Science

There are two contrasting views of the purpose of science and technology. They fit a conflict described by Teilhard de Chardin more than 30 years ago. "What finally divides the men of today into two camps is not class but an attitude of mind—the spirit of movement." He explained: "On the one hand there are those who simply wish to make the world a comfortable dwelling place; on the other hand, those who can only conceive of it as a machine for progress—or better, an organism that is progressing."[6]

Carl Madden says some people see science "as part of culture, linking it with the human urge to comprehend and adapt to the reality of the universe." Others regard it as "a vehicle for industrialization and economic growth." But the rationale that now supports science and technology may indeed change, he says. "Our task . . . is to integrate utility-oriented and knowledge-oriented perceptions of science in order to achieve greater accord between short-term and long-term human welfare."[7]

We are in debt to the scientist, not so much for the products we have derived from his work, but for advancing knowledge on behalf of all of us. The scientific community and industry do themselves an injustice when they defend their work solely on the basis of the products and services they deliver. They should relate to something even more basic than the social problems they are trying to solve. Madden has suggested that it may be more important for business to explain its R&D in terms of the thrust of scientific inquiry than consumer gains. If it can honestly do so, industry might help all of us overcome the stigma of materialism and address ourselves to higher purposes.

Unfortunately, the people who should be defending science and technology on philosophical grounds are not well equipped to do so. The scientific disciplines are not respected as they once were. "People have no link with the history or philosophy of their discipline, and this cheapens the discipline," says Dr. Colodny at the University of Pittsburgh. In math, chemistry, and physics, you once could trace the development of great students and teachers at particular schools. Now, he says, loyalty to the school is gone. "Teachers used to be 'called' to a school; now they are hired. There used to be an implied sense of mission, dedication, and sacrifice."

If we can support only the short-term, product-oriented perception of technological development, we are indeed living off the corpse of science. A corpse cannot humanize or justify our progress. Scientification of our technology offered us a better view of ourselves and of the world. We see more deeply within man, and we are more aware of the global context. Science humanized us; now we are wrestling with the democratization of science. We are more involved than ever.

But man seems to be at a point in his development where he can choose whether he will evolve further. He hesitates. He is tempted to call a halt to the progress that has lifted him far from his crude beginnings.

Man's upright posture, his use of tools, and his development of language each helped the other advance, says Carl Sagan, scientist and author of *The Dragons of Eden.*[8] Chardin wrote earlier: "If the creature from which man issued had not been a biped, his hands would not have been free in time to release the jaws from their prehensile function, and the thick band of maxillary muscles which had imprisoned the cranium could not have been relaxed. It is thanks to two-footedness freeing the hands that the brain was able to grow . . ."[9]

How free are we to reject our further evolution? Chardin warns: "We cannot recapture the animal security of instinct. Because, in becoming man, we have acquired the power of looking to the future and assessing the value of things, we cannot do nothing, since our very refusal to decide is a decision in itself."[10] George Bugliarello of the Polytechnic Institute of New York says, "As a biological process, technology cannot be turned off any more than we can go to our Creator and say, 'Please take back the last hundred cubic centimeters of brain that you gave us some fifty thousand years ago.' "

Evolving A Culture For Modern Man

Man's problem, at this stage of his evolution, is due to his failure to adapt his culture to the world he has created, says Peccei.[11] Fortunately, he says, we don't have to do the impossible—change human nature. We need a cultural evolution—not a biological one. It won't be achieved easily, but it is within our capabilities.[12]

Peccei says the problem is that modern man hasn't learned how to be modern man. Victor Ferkiss expresses the same concept: "Technological man does not yet exist. His job is to invent not the future but first of all himself."[13]

Man does not know himself well, and most of what he has seen of himself in the few thousand years of recorded history reflects only his participation in societies built on competition for food, resources, and power. The "human" weaknesses he sees in the earliest of legends and writings have not been overcome, so he concludes that man is innately aggressive, weak, and deceitful.

If man can now put concerns with nature and the machine behind him, as Bell says, he then confronts himself—a challenge for which he is poorly prepared. For thousands of years man has been divided. He now faces the prospect of having to establish relationships for which he has no precedent. He is unwilling to follow leaders and inexperienced at cooperating voluntarily, says William Glasser in the *Identity Society*.[14]

We stumble because of this fundamental lack of trust in ourselves. There is sufficient evidence from anthropology, however, to show that man is not necessarily aggressive by nature. Anthropologist Richard E. Leakey, for example, asserts that man is not innately inclined either to aggression or peace. He finds man basically a cooperative animal resorting to aggression or territoriality only when important resources are not available.[15]

Science can teach us more about the true nature of man. If we could rightfully shift the blame for our ills from our genes to our culture, we could then work on humanizing our culture. We could change institutions that are built, not on trust, but checks and balances of power. They can be refocused to provide a global, long-term view of man and accommodate the rise of individualism and diversity.

"In the past, the diversity of cultures was balanced by the uniformity of individuals within those cultures," says Jean-Francois Revel. "In the future, cultures will be created by individuals."[16] The process has already begun in the U.S., he believes.

The ascendance of the individual need not undermine society—only society as we know it. Modern man's culture promises to manifest itself as many cultures built on discovery. Diversity need not be divisive, however. Richard Leakey and Roger Lewin write in *Origins* that apparent "deep differences between those cultures should not be seen as divisions between people. Instead, cultures should be interpreted for what they really are: the ultimate declaration of belonging to the human species."[17]

Modern man's culture will be constructed on a positive view of man. His science will not be regarded as an intrusion into forbidden knowledge; his technology will not be an instrument of power to be used by one man against another. This culture will unlock man's potential by tending not only to his animal needs but his inquiring spirit.

Modern man will accept the fire that burns within him, realizing that it is not better or worse than evolution. He will recognize that technology is good or bad depending on the society he fashions and what it asks of technology.

He has just begun to grapple with making technology respond to his needs in global and long-term dimensions. As he makes this awkward start in democratizing technology, however, he could construct a society that extinguishes rather than sustains his fire. Unless he knows how to keep the innovation process functioning, he could kill it before it can make the transition from war to peace, from random advance to rational progress, and from satisfying selfish pursuits to meeting the needs of the planet.

Man will not settle for a future without hope, and hope lies only in change. He must, therefore, make technology his and give it direction. He may tremble as he carries the torch evolution gave him, but he knows the answer to the question "Is it wrong to turn back?"

Chapter Notes

Chapter 1 Fire and Freedom

Page 3. Teilhard de Chardin, "The Directions and Conditions of the Future," (1948) *The Future of Man*, p. 239.

1. John Luther Adams, *Paul Tillich's Philosophy of Culture, Science, and Religion,* pp. 105–106.
2. Pierre Lecomte du Nouy, *Human Destiny*, p. 82 Mentor Book edition.
3. *Ibid.*, p. 89.
4. Lewis M. Branscomb, National Symposium on Technology and Society proceedings.
5. George Bugliarello, National Symposium on Technology and Society proceedings.
6. H. P. Rickman, *Living with Technology*, p. 24.
7. Eugen Loebl, *Humanomics: How We Can Make the Economy Serve Us—Not Destroy Us,* p. 34.
8. *Ibid.*, p. 34.
9. Max Lerner, *America As a Civilization*, vol. 1, p. 226.
10. E. F. Denison, *Accounting for United States' Economic Growth 1929–1969*.
11. U.S. Dept. of Commerce, *Social Indicators 1976*, p. 453.
12. Jean-François Revel, *The Totalitarian Temptation*, p. 194. Used by permission of Doubleday & Company, Inc.
13. Jean-François Revel, *Without Marx or Jesus.*
14. Max Lerner, *op. cit.*, p. 228.
15. Jacob Bronowski, *Science and Human Values,* p. 79.

Chapter 2 Less Innovation, Limited Options

Page 13. Carl Sagan, *The Dragons of Eden*: Speculations on the Evolution of Human Intelligence, p. 236.

1. Arthur Kantrowitz, National Symposium on Technology and Society proceedings.
2. Assistant Secretary for Science and Technology, U.S. Dept. of Commerce, "U.S. Technology Policy"—a draft study, p. 42.
3. National Science Board, *Science Indicators 1974*, p. 100.
4. *Ibid.*, p. 19.
5. J. Herbert Hollomon, National Symposium on Technology and Society proceedings.
6. National Science Board, *op. cit.*, p. 105.
7. *Ibid.*, p. 20.

Chapter 3 Growing List of Wants

1. Arthur J. Scott, "The Food Crisis: A Problem Revisited," *Battelle Today*, November 1977, p. 6.
2. S. H. Wittwer, "The Next Generation of Agricultural Research," *Science*, Jan. 27, 1978, p. 375.
3. Nathan Rosenberg, *Perspectives on Technology*, pp. 242–243.

4. R. Buckminster Fuller, *Utopia or Oblivion: the Prospects for Humanity, p. 5.*
5. "International Economic Report of the President," January 1977, p. 123.
6. Richard E. Leakey and Roger Lewin, *Origins*, p. 61. Reprinted by permission of the publishers, E. P. Dutton.

Chapter 4 Broken Promises

Page 39. Clark C. Abt, *The Social Audit for Management*, p. 4.

1. Teilhard de Chardin, "A Note on Progress," (1920) *The Future of Man*, p. 18.
2. Jean-Francois Revel, *Without Marx or Jesus*, p. 98. Used by permission of Doubleday & Company, Inc.
3. *Ibid.*, p. 98.
4. Hal Hellman, *Technophobia* p. 37 Reprinted by permission of the publishers, M. Evans and Company, Inc.
5. Donald Alstadt, National Symposium on Technology and Society proceedings.
6. Lewis M. Branscomb, National Symposium on Technology and Society proceedings.
7. Jacob Bronowski, *Science and Human Values*, p. 90.
8. H. P. Rickman, *Living with Technology*, p. 89.
9. Fletcher L. Byrom, National Symposium on Technology and Society proceedings.
10. Ben Wattenberg, *The Real America*,p. 300.
11. Clark C. Abt, *op. cit.*, p. 164.

Chapter 5 Too Much Growth

Page 47. Jacob Bronowski, *The Ascent of Man*, p. 438.

1. Paul R. Ehrlich, *The Population Bomb.*
2. Donella H. Meadows et al., *The Limits to Growth–a Report to the Club of Rome.*
3. Aurelio Peccei, *The Human Quality*, p. 85.
4. Mihajlo Mesarovic and Eduard Pestel, *Mankind at the Turning Point*, p. 142. Reprinted by permission of the publishers, E. P. Dutton.
5. Max Lerner, *America As a Civilization*, Vol. 1, pp. 251–252.
6. E. F. Schumacher, *Small Is Beautiful.*

Chapter 6 The Perception of Change

Page 55. Kurt Vonnegut Jr., *Slaughterhouse-Five*, p. 16.

1. For examples, see: *Eureka: An Illustrated History of Inventions From the Wheel to the Computer*, edited by Edward de Bono, pp. 234–237; "The Hundred Major Technological Breakthroughs of All Time," by J. F. Kincaid, *Action*—Journal of the Association for the Advancement of Invention & Innovation, July-August, 1976.
2. Hal Hellman, *Technophobia*, p. 18. Reprinted by permission of the publishers, M. Evans and Company, Inc.
3. *Ibid.*, pp. 24–25.
4. Alvin Toffler, *Future Shock*, p. 285.
5. *Ibid.*, p. 283.
6. P. R. Whitfield, *Creativity in Industry*, p. 200. Reprinted by permission of Penguin Books Ltd.

Chapter 7 Planet in Peril

Page 62. Alfred N. Whitehead, *Science and the Modern World*, p. 208 Mentor Book edition.

1. Charles E. Roth, *The Most Dangerous Animal in the World*, p. 36.

Chapter 8 Inhumane Systems

Page 71. Eldridge Cleaver, *Soul on Ice*, p. 202.

1. P. R. Whitfield, *Creativity in Industry*, p. 188. Reprinted by permission of Penguin Books Ltd.
2. Aurelio Peccei, *The Human Quality*, p. 60.
3. Lyle Schaller, *Understanding Tomorrow*, p. 99.
4. Nathan Rosenberg, *Perspectives on Technology*, p. 216.
5. James Koerner, *Hoffer's America*, p. 59.
6. W. D. McElroy, "The Global Age: Roles of Basic and Applied Research," *Science*, April 15, 1977, p. 268.

Chapter 9 Destroyer of Jobs

Page 77. Frederick Herzberg, *Work and the Nature of Man*, p. 174. Reprinted by permission of Harper & Row, Publishers, Inc.

1. Analyses of 500 largest industrial corporations, *Fortune*, July 1967 and May 1977.
2. Assistant Secretary for Science and Technology, U.S. Dept. of Commerce, "U.S. Technology Policy"—a draft study, p. 21.
3. Frederick Herzberg, *op. cit.*, Chapter 6.
4. Frederick Herzberg, "The end of obligation," *Industry Week*, Oct. 16, 1972, p. 60.
5. William Glasser, *The Identity Society*, p. 9.
6. Roger D'Aprix, *In Search of a Corporate Soul*, p. 158.
7. "Workstations enhance jobs at Citibank," *Industry Week*, Oct. 10, 1977, p. 40.
8. "Can you win?" *Industry Week*, Jan. 23, 1978, p. 76.
9. *Ibid.*, p. 76.
10. Hazel Henderson, "The Mirage of Efficiency," *WPI Journal*, Worcester Polytechnic Institute, December 1976, p. 17.
11. William Glasser, *op. cit.*, pp. 34–35.

Chapter 10 Overreaction by Overregulation

Page 89. Eric Hoffer, *The True Believer*, p. 115.

1. Sumner Myers and Eldon E. Sweezy, Denver Research Institute, "Federal Incentives for Regulation," p. 29.
2. Robert DeFina, Center for the Study of American Business, "Public Expenditures for Federal Regulation of Business."
3. "Solar energy for office and plant starts to heat up," Industry Week, Oct. 10, 1977, p. 66.
4. Carl H. Madden, "2002," *Across the Board,* October 1976, p. 19.
5. W. Allen Wallis, National Symposium on Technology and Society proceedings.
6. J. Herbert Hollomon, National Symposium on Technology and Society proceedings.

Chapter 11 Flickering Corporate Spirit

Page 101. Warren Bennis, *The Unconscious Conspiracy*, pp. 153–154.

1. Ernest V. Heyn, *Fire of Genius*, p. 15–16.
2. National Science Board, *Science Indicators 1974*, p. 87.
3. Daniel V. De Simone, "Technological Innovation: Its Environment and Management," p. 9.
4. Thomas W. Harvey, "Technical ventures—catalysts for economic growth," *Battelle Today*, August 1977, p. 5.
5. Nathan Rosenberg, *Perspectives on Technology*, p. 285.
6. Edwin A. Gee and Chaplin Tyler, *Managing Innovation*, pp. 172–173.
7. Curt Nicolin, *Private Industry in a Public World*, p. 67. Reprinted by permission of Addison-Wesley Publishing Co.
8. *Ibid*.

Chapter 12 The People Will Decide

Page 119. Reinhold Niebuhr, *The Children of Light and the Children of Darkness*, p. 118.

1. James B. Conant, *Modern Science and Modern Man*, p. 66.
2. *Ibid*., p. 67.
3. Robert Heilbroner, *Business Civilization in Decline*, pp. 57--58.
4. James Koerner, *Hoffer's America*, p. 133.
5. Philip Lesly, *The People Factor*, p. 61.
6. *Ibid*., pp. 63–64.
7. Jean-François Revel, *Without Marx or Jesus*, p. 104. Used by permission of Doubleday & Company, Inc.
8. *Ibid*., p. 172.
9. Alvin Toffler, *The Eco-Spasm Report*, p. 101.
10. Aurelio Peccei, *The Human Quality*,pp. 177.
11. John W. Gardner, *The Recovery of Confidence*, pp. 72–73.
12. E. F. Schumacher, *Small Is Beautiful*, p. 70.
13. Aurelio Peccei, *op. cit*., p. 177.
14. *Ibid*., p. 22.
15. Victor C. Ferkiss, *Technological Man: The Myth and the Reality,* p. 137.
16. Edward E. David Jr., National Symposium on Technology and Society proceedings.
17. Jean-Francois Revel, *The Totalitarian Temptation,* p. 39. Used by permission of Doubleday & Company, Inc.
18. John W. Gardner, *op. cit*., p. 131.
19. William Glasser, *The Identity Society*, p. 235.
20. Aurelio Peccei, *op. cit*., p. 96.
21. Daniel Bell, *The Coming of Post-Industrial Society; A Venture in Social Forecasting,* p. 488.
22. Alvin Toffler, *Future Shock*, p. 167.
23. Werner J. Dannhauser, "What Constitutes 'Quality' in Life?" *Qualities of Life*, p. 173.

Chapter 13 Private Enterprise and Public Goals

Page 133. J. Herbert Hollomon, National Symposium on Technology and Society proceedings.

1. Nathan Rosenberg, *Perspectives on Technology*, pp. 126–127.
2. Donald Alstadt, personal interview and National Symposium on Technology and Society keynote speech.
3. C. Lester Hogan, National Symposium on Technology and Society proceedings.
4. Robert Heilbroner, *Business Civilization in Decline*, pp. 23–25.
5. Aurelio Peccei, *The Human Quality*, p. 46.
6. Ian H. Wilson, "Business and the Future: Social Challenge, Corporate Response," *The Next 25 Years*, edited by Andrew A. Spekke, p. 143.
7. George S. Dominguez, *Business, Government and the Public Interest,* p. 40.
8. J. Herbert Hollomon, *op. cit.*
9. Jerome Wiesner, testimony before the Subcommittee on Economic Growth of the Joint Economic Committee of Congress, July 15–16, 1975.
10. Assistant Secretary for Science and Technology, U. S. Dept. of Commerce, "U.S. Technology Policy"—a draft study, p. 3.
11. E. F. Schumacher, *Small Is Beautiful*, p. 42.
12. Daniel V. De Simone, "Technological Innovation: Its Environment and Management," p. 12.
13. Roland N. McKean, Center for the Study of American Business, "The Regulation of Chemicals and the Production of Information," p. 46.
14. Robert Gilpin, testimony before the Subcommittee on Economic Growth of the Joint Economic Committee of Congress, July 15–16, 1975.

Chapter 14 Dialogue and Control

Page 147. George Bugliarello, National Symposium on Technology and Society proceedings.

1. Clark C. Abt, *The Social Audit for Management*, p. 164.
2. Arthur Kantrowitz, National Symposium on Technology and Society proceedings.
3. Clyde Z. Nunn, "Is There a Crisis of Confidence in Science?" *Science*, Dec. 9. 1977, p. 995.
4. Gerard Piel, *Science in the Cause of Man*, p. 75.
5. June Goodfield, "Humanity in Science: A Perspective and a Plea," *Science*, Nov. 11, 1977, p. 585.
6. *Ibid.*, p. 583.
7. Richard M. Restak, *Premeditated Man,* pp. 162—163 Penguin Books edition.
8. Robert M. May, "The Recombinant DNA Debate," *Science*, Dec. 16, 1977, p. 1145.
9. Carl H. Madden, *Clash of Culture: Management in an Age of Changing Values*. p. 109.
10. June Goodfield, *op. cit.*, p. 583.

Chapter 15 If Man Is to Be Man

Page 157. Robert Bridges, *The Testament of Beauty*, lines 339–343.

1. Gerard Piel, *Science in the Cause of Man*, p. 111.
2. Daniel Bell, *The Coming of Post-Industrial Society; A Venture in Social Forecasting*, p. 45.
3. Jacob Bronowski, *The Ascent of Man*, p. 437.
4. Jacob Bronowski, *Science and Human Values*, p. 90.
5. *Ibid.*, p. 93.
6. Teilhard de Chardin, "A Great Event Foreshadowed: The Planetisation of Mankind," (1945) *The Future of Man*, p. 144.
7. Carl H. Madden, *Clash of Culture: Management in an Age of Changing Values*, p. 31.
8. Carl Sagan, *The Dragons of Eden: Speculations on the Evolution of Human Intelligence*, p. 173.
9. Teilhard de Chardin, *The Phenomenon of Man*, p. 170.
10. Teilhard de Chardin, "The Grand Option," (1939) *The Future of Man*, p. 49.
11. Aurelio Peccei, *The Human Quality,* p. xi.
12. *Ibid.*, pp. 31–32.
13. Victor C. Ferkiss, *Technological Man: The Myth and the Reality,* p. 201.
14. William Glasser, *The Identity Society*, p. 229.
15. Richard E. Leakey and Roger Lewin, *Origins*, p. 213.
16. Jean-François Revel, *Without Marx or Jesus*, p. 74. Used by permission of Doubleday & Company, Inc.
17. Leakey and Lewin, *op. cit.*, p. 256. Reprinted by permission of the publishers, E. P. Dutton.

Bibliography

Abt, Clark C., *The Social Audit for Management.* New York: AMACOM, a division of American Management Associations, 1977.

Adams, James Luther, *Paul Tillich's Philosophy of Culture, Science, and Religion.* New York: Harper & Row, Publishers, Inc., 1965.

American Society of Mechanical Engineers, New York: *Technology and Society Newsletter,* Fall 1977.

Assistant Secretary for Science and Technology, U.S. Dept. Of Commerce, "U.S. Technology Policy"—a draft study, Washington, March 1977.

Backman, Jules, ed., *Business and the American Economy, 1976–2001.* New York: New York University Press, 1976.

Beckerman, Wilfred, *The Cheers for the Affluent Society.* New York: Saint Martin's Press, 1974.

Bell, Daniel, *The Coming of Post-Industrial Society.* A Venture in Social Forecasting. New York: Basic Books Inc., 1973.

Bennis, Warren, *The Unconscious Conspiracy.* New York: AMACOM, a division of American Management Associations, 1976.

Bridges, Robert, *The Testament of Beauty.* Oxford: Oxford University Press, 1929.

Bronowski, Jacob, *The Ascent of Man.* Boston: Little, Brown and Company, 1973.

Bronowski, Jacob, *Science and Human Values.* New York: Julian Messner Inc., 1956.

Budget of the United States Government–Fiscal Year 1979. Washington: U.S. Government Printing Office, 1978.

Cleaver, Eldridge, *Soul on Ice.* New York: McGraw-Hill Book Company, 1968.

Conant, James B., *Modern Science and Modern Man.* New York: Columbia University Press, 1952.

Dannhauser, Werner J., "What Constitutes 'Quality' in Life?" *Qualities of Life* (papers prepared for the Commission on Critical Choices for Americans). Lexington, Massachusetts: D. C. Heath & Co., 1976.

D'Aprix, Roger M., *In Search of a Corporate Soul.* New York: AMACOM, a division of American Management Associations, 1976.

de Bono, Edward, and the editors of the London *Sunday Times, Eureka! An Illustrated History of Inventions from the Wheel to the Computer.* London: Thames and Hudson Ltd., 1974.

DeFina, Robert, "Public and Private Expenditures for Federal Regulation of Business." Center for the Study of American Business, Washington University, St. Louis, November 1977.

Denison, E. F., *Accounting for United States' Economic Growth 1929–1969.* Washington: Brookings Institution, 1974.

De Simone, Daniel V., "Technological Innovation: Its Environment and Management." Report by Panel on Invention and Innovation convened by the Secretary of Commerce, January 1967.

Dominguez, George S., *Business, Government, and the Public Interest.* New York: AMACOM, a division of American Management Associations, 1976.

Economic Report of the President–January 1978. Washington: U.S. Government Printing Office, 1978.

Ehrlich, Paul R., *The Population Bomb.* New York: The Sierra Club, 1969.

Ferkiss, Victor C., *Technological Man: The Myth and the Reality*. New York: George Braziller Inc., 1969. Mentor Book, 1970.

Fortune, July 1967 and May 1977 analyses of 500 largest industrial corporations.

Fuller, R. Buckminster, *Utopia or Oblivion: the Prospects for Humanity*. New York: Overlook Press, 1969.

Gardner, John W., *The Recovery of Confidence*. New York: W. W. Norton & Company Inc., 1970.

Gartmann, Heinz, *Rings Around the World*. New York: William Morrow & Co.,1959.

Gee, Edwin A., and Tyler, Chaplin, *Managing Innovation*. New York: John Wiley & Sons Inc., 1976.

Glasser, William, *The Identity Society*. New York: Harper & Row, Publishers, Inc., 1972.

Goodfield, June, "Humanity in Science: A Perspective and a Plea," *Science,* Vol. 198, pp. 580–585, November 11, 1977.© 1977 by the American Association for the Advancement of Science.

Gould Inc., Rolling Meadows, Illinois: "Technology." A series of "white papers."

Gray, Elizabeth; Gray, David Dodson; and Martin, William F., *Growth and Its Implications for the Future*. Branford, Connecticut: The Dinosaur Press, 1975.

Gyllenhammar, Pehr, *People at Work*. Reading, Massachusetts: Addison-Wesley Publishing Co., 1977.

Harvey, Thomas W., "Technical ventures—catalysts for economic growth," *Battelle Today*, Battelle Memorial Institute, Columbus, Ohio, August 1977.

Heilbroner, Robert I., *Business Civilization in Decline*. New York: W. W. Norton & Company Inc., 1976.

Hellman, Hal, *Technophobia*. New York: M. Evans and Company Inc., 1976.

Henderson, Hazel, "The Mirage of Efficiency," *WPI Journal*. Worcester Polytechnic Institute, Worcester, Massachusetts, December 1976.

Herzberg, Frederick, *Work and the Nature of Man*. Cleveland: World Publishing Co., 1966.

Herzberg, Frederick, "The end of obligation," *Industry Week,* Oct. 16, 1972.

Heyn, Ernest V., *Fire of Genius*. Garden City, New York: Anchor Press/Doubleday, 1976.

Hoffer, Eric, *The True Believer*. New York: Harper & Row, Publishers, Inc. 1951.

Industry Week, "Solar energy for office and plant starts to heat up," October 10, 1977.

Industry Week, "Workstations enhance jobs at Citibank," October 10, 1977.

Industry Week, "Can you win?" January 23, 1978.

International Economic Report of the President–January 1977. Washington: U.S. Government Printing Office, 1977.

Joint Economic Committee, Congress of the United States, "Hearings before the Subcommittee on Economic Growth," July 15–16, 1975.

Junker, Louis, ed., *The Political Economy of Food and Energy*. Ann Arbor: The University of Michigan Press, 1977.

Kahn, Herman; Brown, William; and Martel, Leon, *The Next 200 Years*. New York: William Morrow & Company Inc., 1976.

Kincaid, J. F., "The Hundred Major Technological Breakthroughs of All Time," *Action,* Journal of the Association for the Advancement of Invention & Innovation. Arlington, Virginia, July–August 1976.

Koerner, James D., *Hoffer's America*. LaSalle, Illinois: Open Court Publishing Co., 1973.

Leakey, Richard E., and Lewin, Roger, *Origins*. New York: E. P. Dutton & Co. Inc., 1977.

Lecomte du Nouy, Pierre, *Human Destiny*. London: Longmans, Green & Co., Inc., 1947.

Lerner, Max, *The Age of Overkill*. New York: Simon and Schuster, 1962.

Lerner, Max, *America As a Civilization*, Vol. 1. New York: Simon and Schuster, 1957.

Lesly, Philip, *The People Factor*. Homewood, Illinois: Dow Jones—Irwin Inc., 1974.

Loebl, Eugen, *Humanomics: How We Can Make the Economy Serve Us–Not Destroy Us*. New York: Random House, Inc., 1976.

McElroy, W. D., "The Global Age: Roles of Basic and Applied Research," *Science*, Vol. 196, pp. 267–270, April 13, 1977.© 1977 by the American Association for the Advancement of Science.

McKean, Roland N., "The Regulation of Chemicals and the Production of Information," Center for the Study of American Business, Washington University, St. Louis, June 1976.

Madden, Carl H., *Clash of Culture: Management in an Age of Changing Values*. Washington: National Planning Association, 1972.

Madden, Carl H., "2008," *Across the Board*, The Conference Board Inc., New York, October 1976.

May, Robert M., "The Recombinant DNA Debate," *Science* Vol. 198, pp. 1144–1145, December 16, 1977.© 1977 by the American Association for the Advancement of Science.

Meadows, Donella H.; Meadows, Dennis L.; Randers, Jorgen; Behrens, William W. III, *The Limits to Growth–a Report to the Club of Rome*. New York: Universe Books, 1972.

Mesarovic, Mihajlo, and Pestel, Eduard, *Mankind at the Turning Point*. New York: E. P. Dutton & Co. Inc. /Reader's Digest Press, 1974.

Myers, Sumner, and Sweezy, Eldon E., "Federal Incentives for Innovation," Denver Research Institute, University of Denver, January 1976.

National Science Board, National Science Foundation, *Science at the Bicentennial*. Washington: U.S. Govt. Printing Office, 1976.

National Science Board, National Science Foundation, *Science Indicators 1974*. Washington: U.S. Govt. Printing Office, 1975.

National Symposium on Technology an Society, October 3–4, 1977, Proceedings. Erie, Pennsylvania: Lord Corp., 1978.

Nicolin, Curt, *Private Industry in a Public World*. Reading, Massachusetts: Addison-Wesley Publishing Co., 1977.

Niebuhr, Reinhold, *The Children of Light and the Children of Darkness*. New York: Charles Scribner's Sons, 1944.

Nunn, Clyde Z., "Is There a Crisis of Confidence in Science?" *Science*, Vol. 198, p. 995, December 9, 1977.© 1977 by the American Association for the Advancement of Science.

Oliver, John W., *History of American Technology*. New York: The Ronald Press Company, 1956.

Orwell, George, *Nineteen Eighty-Four*. New York: Harcourt, Brace & World Inc., 1949.

Peccei, Aurelio, *The Human Quality*. Oxford: Pergamon Press, 1977.

Piel, Gerard, *Science in the Cause of Man*. New York: Alfred A. Knopf Inc., 1961.
Restak, Richard M., *Premeditated Man*. New York: The Viking Press Inc., 1975. Penguin Books, 1977.
Revel, Jean-François, *The Totalitarian Temptation*. Garden City, New York: Doubleday & Company Inc., 1977.
Revel, Jean-François, *Without Marx or Jesus*. Garden City, New York: Doubleday & Company Inc., 1971.
Rickman, H. P., *Living With Technology*. London: Hodder and Stoughton Ltd., 1967.
Rosenberg, Nathan, *Perspectives on Technology*. Cambridge: Cambridge University Press, 1976.
Roth, Charles E., *The Most Dangerous Animal in the World*. Reading, Massachusetts: Addison-Wesley Publishing Co., 1971.
Sagan, Carl, *The Dragons of Eden: Speculations on the Evolution of Human Intelligence* New York: Random House, Inc. 1977.
Schaller, Lyle E., *Understanding Tomorrow*. Nashville, Tennessee: Abingdon Press, 1976.
Schumacher, E. F., *Small Is Beautiful*. New York: Harper & Row, Publishers, Inc. 1973.
Scott, Arthur J., "The. Food Crisis: A Problem Revisited," *Battelle Today*, Battelle Memorial Institute, Columbus, Ohio, November 1977.
Spekke, Andrew A., ed., *The Next 25 Years, Crisis and Opportunity*. Washington: World Future Society, 1975.
Teilhard de Chardin, Pierre, *The Future of Man*. New York: Harper & Row, Publishers, Inc., 1964.
Teilhard de Chardin, Pierre, *The Phenomenon of Man*. New York: Harper & Row, Publishers, Inc., 1959.
Toffler, Alvin, *The Eco-Spasm Report*. New York: Bantam Books Inc., 1975.
Toffler, Alvin, *Future Shock*. New York: Random House, Inc., 1970.
U.S. Dept. of Commerce, *Social Indicators 1976*. Washington: U.S. Government Printing Office, 1977.
Vacca, Roberto, *The Coming Dark Age*. Garden City: Doubleday & Company, Inc., 1973.
Vonnegut, Kurt Jr., *Slaughterhouse-Five*. New York: Delacorte Press /Seymour Lawrence, 1969.
Wattenburg, Ben J., *The Real America*. Garden City, New York: Doubleday & Company, Inc., 1974.
Weber, James A., *Grow or Die!* New Rochelle: Arlington House, 1977.
Whitehead, Alfred North, *Science and the Modern World*. New York: The Macmillan Company, 1925. Mentor Book, 1948.
Whitfield P. R., *Creativity in Industry*. Harmondsworth, Middlesex: Penguin Books Ltd., 1975.
Wittwer, S. H., "The Nex Generation of Agricultural Research," *Science*, Vol. 199, p. 375, January 27, 1978. © 1978 by the American Association for the Advancement of Science.